Strive for a 5:

Preparing for the AP* Environmental Science Examination

to accompany

Friedland & Relyea
*Environmental Science for AP**

Courtney Mayer

W. H. Freeman and Company
New York

ISBN-13: 978-1-464-10869-3
ISBN-10: 1-464-10869-2

Printed in the United States of America

Fourth Printing

W.H. Freeman and Company
41 Madison Avenue
New York, NY 10010
Houndmills, Basingstoke RG21 6XS, England

www.whfreeman.com

TABLE OF CONTENTS

PREFACE

This book, *Strive for a 5: Preparing for the AP* Environmental Science Examination*, is designed for use with *Friedland and Relyea's Environmental Science for AP**. It is intended to help you evaluate your understanding of the material covered in the textbook, reinforce the key concepts you need to learn, and prepare you to take the AP Environmental Science exam. It is a good idea to read through this guide early in the course so that you have a solid understanding of what you are preparing for from the start.

This book is divided into two sections: a study guide section and a test preparation section.

THE STUDY GUIDE SECTION

The study guide section is designed for you to use throughout your AP course. As each module is covered in your class, you can use the study guide to help you identify and reinforce the important concepts in environmental science.

For each chapter, the study guide is organized as follows:
Summary: An overview of the chapter
While You Read: Key ideas and vocabulary terms for each subsection of the chapter
Chapter Review: A summary of the most important concepts in the chapter
After You Read: Short answer questions to see if you understood what you read
Do the Math: Practice problems to help you with the math that will be on the exam

An answer key is found in the back of the book so that you can check your work on the "While You Read," "After You Read" and "Do the Math" problems.

THE TEST PREPARATION SECTION

So that you can see what it's like to take a full, timed exam, this guide features two full-length practice exams. After completing the exams, you can check the answers in the back. Be sure to look closely at the questions that you answered incorrectly and spend extra time with the concepts found in those questions.

Best of luck on the AP exam!

Courtney Mayer

Chapter 1: Studying the State of Our Earth

Summary

This chapter introduces the concept and importance of environmental science. It explains the five environmental indicators that we use to judge the health of the planet. The scientific method is also explained.

While You Read

Environmental Science offers important insights into our world and how we influence it

Key Ideas

- Our environment includes the living and nonliving components of our planet.
- Environmental science uses knowledge from many disciplines.

Match These Key Terms

_____1. Environment

_____2. Environmental Science

_____3. System

_____4. Ecosystem

_____5. Biotic

_____6. Abiotic

_____7. Environmentalist

_____8. Environmental Studies

a. Field of study that includes environmental science, environmental policy, economics, literature, and ethics

b. Living

c. Nonliving

d. A person that seeks to protect the environment

e. A particular location on Earth distinguished by its mix of interacting biotic and abiotic components

f. An interacting set of components that influence one another by exchanging energy or materials

g. The field of study that looks at interactions among human systems and those found in nature

h. The sum of all the conditions surrounding us that influence life

Humans alter natural systems

Key Ideas

- Humans have been altering the Earth for millions of years.
- Technology and the growing human environment have had a global environmental impact.

Environmental Scientists monitor natural systems for signs of stress

Key Ideas

- The natural world provides many things for us, such as clean water, timber, fisheries, and crops.
- Genetics, species, and ecosystem diversity are the different forms of biodiversity.

- The five global indicators that help us monitor the health of our ecosystem are biodiversity, food production, global surface temperature, the size of the human population and resource depletion.

Match These Key Terms

_____1. Ecosystem services

_____2. Environmental indicators

_____3. Sustainability

_____4. Biodiversity

_____5. Species

_____6. Speciation

_____7. Background extinction rate

_____8. Greenhouse gases

_____9. Anthropogenic

_____10. Development

a. The diversity of life forms in an environment

b. The process by which natural environments provide life-supporting resources.

c. Living on Earth in a way that allows humans to use its resources without depriving future generations of those resources

d. A group of organisms that is distinct from other groups in its morphology, behavior, or biochemical properties

e. The evolution of new species

f. An indicator that describes the current state of an environmental system

g. Gasses in Earth's atmosphere that trap heat near the surface

h. Improvement in human well-being through economic advancement

i. The average rate at which species become extinct over the long term

j. Derived from human activities

Human well-being depends on sustainable practices

Key Ideas

- Living sustainably means acting in a way so that activities that are crucial to human society can continue.
- Basic human needs include air, water, food, and shelter.
- A person's ecological footprint is a measure of how much that person consumes, and is expressed in area of land.

Fill-in-the-Blank

1. E.O. Wilson wrote that humans exhibit biophilia, which is the love of _____.

2. One way to assess if we are living sustainably is to measure a person's _____

   _____.

Science is a process

Key Ideas

- The steps of the scientific method are: observe and question, form a testable hypothesis, collect data, interpret results, and disseminate findings.

- In an experiment, the group that you are testing is called the experimental group.
- A control group is a group that experiences exactly the same conditions as the experimental group, except for the single variable under study.

Match These Key Terms

_____1. Scientific method

_____2. Hypothesis

_____3. Null hypothesis

_____4. Replication

_____5. Sample size

_____6. Accuracy

_____7. Precision

_____8. Uncertainty

_____9. Inductive reasoning

_____10. Deductive reasoning

_____11. Critical thinking

_____12. Theory

_____13. Natural law

_____14. Control group

_____15. Natural experiment

a. The process of questioning the source of information, considering the methods used to obtain the information, and drawing conclusions

b. The process of applying a general statement to specific facts or situations

c. A statement or idea that can be falsified, or proven wrong

d. An objective method to explore the natural world, draw inferences from it, and predict the outcome of certain events, processes, or changes

e. An estimate of how much a measured or calculated value differs from a true value

f. The data collection procedure of taking repeated measurements

g. In a scientific investigation, a group that experiences exactly the same conditions as the experiment group, except for the single variable under study

h. A hypothesis that has been repeatedly tested and confirmed by multiple groups of researchers and has reached wide acceptance

i. How close a measured value is to the actual or true value

j. A theory for which there is no known exception and that has withstood rigorous testing

k. A natural event that acts as an experimental treatment in an ecosystem

l. The process of making general statements from specific facts or examples

m. The number of times a measurement is replicated in the data collection process

n. A testable theory or supposition about how something works

o. How close the repeated measurements of a sample are to one another

Environmental science presents unique challenges

Key Ideas

- Challenges to environmental science include:
  - The lack of a "control" planet to which Earth can be compared.

- o The fact that many dilemmas are subjective—there is no single measure of environmental quality.
- o The number of interacting systems in environmental science makes it a very complex science and one that is frequently subject to debate.

Fill-in-the-Blank

1. Environmental justice is a field of study that works toward _____ enforcement of environmental laws.

 Chapter Review

There will be few direct questions on the AP exam from this chapter. However, it is very important for you to learn the key terms because they introduce information that will be covered in future chapters. An important idea from this chapter is the scientific method—the AP exam frequently includes questions that test your understanding of how a proper experiment is designed and your familiarity with terms like "control group" or "natural experiment."

List questions from your initial reading of the chapter

_____

_____

_____

_____

_____

_____

_____

_____

 **After You Read**

Short Answer

1. What disciplines are incorporated into the study of environmental science?

2. List the 5 global-scale environmental indicators.

3. Describe the following: genetic diversity, species diversity, ecosystem diversity.

4. Give an example of an activity that is anthropogenic.

5. Currently, what is the size of the human population?

6. What is a person's ecological footprint?

7. List the steps in the scientific method.

Do The Math

There are 2.47 acres per hectare. Therefore, 1 acre= 0.40 ha.

Convert the following from acres to hectares.

50,000 acres = _____ hectares

75,000 acres = _____ hectares

150,000 acres = _____ hectares

Chapter 2: Environmental Systems

Summary
Chapter 2 is a review of general science concepts. It covers topics such as atoms and molecules from chemistry, cells from biology, and energy topics from physics. Positive and negative feedback loops are addressed, which are important for you to understand.

 **While You Read**

Earth is a single interconnected system
Key Ideas
- The activities of humans, the lives of other organisms, and processes in the environment are all interconnected.

All environmental systems consist of matter
Key Ideas
- All matter is composed of atoms.
- The three types of chemical bonds are covalent, ionic and hydrogen.
- Properties of water include surface tension, capillary action, boiling and freezing, and water as a solvent.
- Carbohydrates, proteins, nucleic acids, and lipids are the basis of biological molecules.

Match These Key Terms

_____1. Matter
_____2. Mass
_____3. Atom
_____4. Element

_____5. Periodic table

_____6. Molecules

_____7. Compounds

_____8. Atomic number
_____9. Mass number

_____10. Isotopes

_____11. Radioactive decay

_____12. Half-life

_____13. Covalent bonds

a. A chemical bond between two oppositely charged ions
b. A substance that contributes hydroxide ions to a solution
c. The number of protons in the nucleus of a particular element
d. The smallest particle that can contain the chemical properties of an element
e. A property of water that results from the cohesion of water molecules at the surface of a body of water and creates a sort of skin on the water's surface
f. The number indicating the strength of acids and bases on a scale from 0 to 14, where 7 is neutral, a value below 7 is acidic, and a value above 7 is basic
g. The spontaneous release of material from the nucleus of radioactive isotopes
h. A substance that contributes hydrogen ions to a solution
i. A substance composed of atoms that cannot be broken down into smaller, simpler components
j. A law of nature stating that matter cannot be created or destroyed
k. A measurement of the total number of protons and neutrons in an element
l. A nucleic acid that contains the genetic material to pass to offspring and contains the code for reproducing the components of the next generation
m. Anything that occupies space and has mass

_____14. Ionic bond

_____15. Hydrogen bonds

_____16. Polar molecule

_____17. Surface tension
_____18. Capillary Action

_____19. Acid

_____20. Base
_____21. pH

_____22. Chemical reaction

_____23. Law of conservation of matter

_____24. Inorganic compounds
_____25. Organic compounds

_____26. Carbohydrates

_____27. Proteins

_____28. Nucleic acids

_____29. DNA
_____30. RNA

_____31. Lipids
_____32. Cell

n. A compound that contains carbon-carbon and carbon-hydrogen bonds
o. A measurement of the amount of matter an object contains
p. A compound that does not contain the element carbon or contains carbon bound to elements other than hydrogen
q. Particles containing more than one atom
r. A chart of all chemical elements currently known, organized by their properties
s. A highly organized living entity that consists of the four types of macromolecules and other substances in a watery solution, surrounded by a membrane
t. The bond formed when elements share electrons
u. Smaller organic biological molecules that do not mix with water
v. A molecule in which one side is more positive and the other side is more negative
w. Long chains of nitrogen-containing organic molecules known as amino acids, critical to living organisms for structural support, energy storage, internal transport, and defense against foreign substances
x. A nucleic acid that translates the code stored in DNA and allows for the synthesis of proteins
y. Atoms of the same element with different numbers of neutrons
z. A weak chemical bond that forms when hydrogen atoms that are covalently bonded to one atom are attracted to another atom on another molecule
aa. A property of water that occurs when adhesion of water molecules to a surface is stronger than cohesion between the molecules
bb. Organic compounds found in all living cells, which form in long chains to make DNA and RNA
cc. The time it takes for one-half of an original radioactive parent atom to decay
dd. Molecules containing more than one element
ee. A reaction that occurs when atoms separate from molecules or recombine with other molecules
ff. Compounds composed of carbon, hydrogen, and oxygen atoms

Energy is a fundamental component of environmental systems

Key Ideas

- Earth's systems cannot function, and organisms cannot survive, without energy.
- Just as matter can neither be created nor destroyed, energy is neither created nor destroyed.
- When energy is transformed, the quantity of energy remains the same, but its ability to do work diminishes.

Match These Key Terms

_____1. Energy
_____2. Electromagnetic radiation

_____3. Photons
_____4. Joule
_____5. Power
_____6. Potential energy

_____7. Kinetic energy

_____8. Chemical energy

_____9. Temperature

_____10. First law of thermodynamics
_____11. Second law of thermodynamics
_____12. Energy efficiency

_____13. Energy quality
_____14. Entropy

a. The energy of motion
b. A law of nature stating that energy can neither be created nor destroyed
c. Randomness in a system
d. The measure of the average kinetic energy of a substance
e. The ability to do work or transfer heat
f. The ratio of the amount of work done to the total amount of energy introduced to the system
g. A form of energy emitted by the Sun that includes, but is not limited to, visible light, ultraviolet light, and infrared energy
h. The law stating that when energy is transformed, the quantity of energy remains the same, but its ability to do work diminishes
i. Massless packets of energy that carry electromagnetic radiation at the speed of light
j. The ease with which an energy source can be used for work
k. The rate at which work is done
l. The amount of energy used when a one-watt electrical device is turned on for one second
m. Stored energy that has not been released
n. Potential energy stored in chemical bonds

Energy conversions underlie all ecological processes

Key Ideas

- Individual organisms rely on a continuous input of energy in order to survive, grow, and reproduce.

Systems analysis shows how matter and energy flow in the environment

Key Ideas

- Environmental scientists look at the whole picture, not just the individual parts of a system, in order to understand how that system works.
- A positive feedback loop amplifies changes, whereas a negative feedback resists changes.

Match These Key Terms

_____1. Open system
_____2. Closed system

_____3. Inputs
_____4. Outputs
_____5. Systems analysis

_____6. Steady state

_____7. Feedback
_____8. Negative feedback loops

_____9. Positive feedback loops

a. An adjustment in input or output rates caused by changes to a system
b. A system in which matter and energy exchanges do not occur across boundaries
c. A feedback loop in which change in a system is amplified
d. Additions to a system
e. A system in which exchanges of matter or energy occur across system boundaries
f. A state in which inputs equal outputs, so that the system is not changing over time
g. Losses from a system
h. Feedback loops in which a system responds to a change by returning to its original state, or by decreasing the rate at which the change is occurring
i. An analysis to determine inputs, outputs, and changes in a system under various conditions

Natural systems change across space and over time

Key Ideas

- Differences in environmental conditions affect what grows or lives in an area.
- Natural systems are affected by the passage of time.

 Chapter Review

This chapter is similar to chapter 1 in that you will not be asked many specific questions from this chapter on the AP exam, but you should know the information so that you can understand concepts in future chapters. Feedback loops are one of the most important concepts in this chapter; make sure you are familiar with the difference between negative and positive feedback loops.

List questions from your initial reading of the chapter

_____

_____

_____

_____

_____

_____

_____

 **After You Read**

Short Answer

1. Briefly explain radioactive decay.

2. What is an element's half-life?

3. List some properties of water and describe how these properties are important to living systems.

4. If a substance has a pH of 3, how many more times acidic is it than a substance with a pH of 5?

5. Give an example of potential energy and of kinetic energy.

6. Explain the first and second laws of thermodynamics in your own words.

7. What is the energy efficiency of a coal burning power plant, an incandescent light bulb and the electrical transmission lines between the power plant and the house?

8. Give an example of a negative and a positive feedback loop.

Do The Math
A family is looking to see how much of their electricity bill is coming from their clothes washer and drier. The washer uses 1500 watts and the clothes drier uses 2000 watts and the family pays $.10 per kilowatt-hour. Each appliance runs approximately 30 minutes each day. How much does the family spend per week to run both these appliances?

Review Practice Questions: Chapters 1-2

Introduction

1. Which of the following is an example of a biotic component of an ecosystem?
   a. Soil
   b. Water
   c. Nitrogen
   d. Sunflower
   e. Rock

2. As a solid or liquid, water has its lowest density at
   a. 0° Celsius.
   b. 32° Celsius.
   c. 100° Fahrenheit.
   d. 4° Celsius.
   e. 100° Celsius.

3. The earth is regulated by feedback loops. These loops are:
   a. positive
   b. negative
   c. both positive and negative
   d. getting weaker
   e. in steady states

4. 7 square miles is equal to _____ acres. (1 square mile = 640 acres)
   a. .4270
   b. 4.270
   c. 42.70
   d. 427.0
   e. 4,720

5. Which of the following is NOT an example of an anthropogenic factor?
   a. Burning of fossil fuels for electricity
   b. Burning of fossil fuels in vehicles
   c. Volcanic eruptions
   d. Deforestation for planting crops
   e. Over-harvesting of ocean resources

Use the graph to answer questions 6-8.

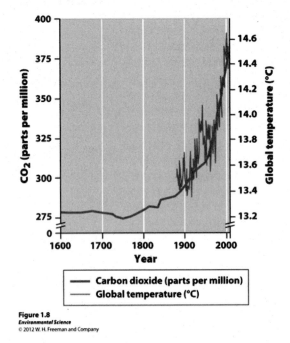

Figure 1.8
Environmental Science
© 2012 W. H. Freeman and Company

6. When carbon dioxide levels were at 325 ppm, what was the approximate global temperature in degrees Celsius?
   a. 1600
   b. 14.1
   c. 13.5
   d. 13.8
   e. 1920

7. If carbon dioxide levels hit 400 ppm, what do you estimate global temperature to be?
   a. 14.8° Celsius
   b. 13.8° Celsius
   c. 17.2° Celsius
   d. 2,100° Celsius
   e. 2,500° Celsius

8. If the trend in global surface temperatures continues, what year do you estimate temperatures to become 15.0° Celsius?
   a. 2000
   b. 2100
   c. 2500
   d. 3000
   e. 3500

9. Greenhouse gases in the atmosphere help
    a. keep UV light from reaching Earth.
    b. regulate temperatures near Earth's surface.
    c. heat to be released back to space.
    d. keep the ozone layer intact.
    e. the earth to stay cooler.

10. The population of the United States is 307,000,000. If Americans consume approximately
    50 billion bottles of water each year, how many bottles is that per capita?
    a. 1
    b. 7
    c. 50
    d. 100
    e. 160

11. If there are 364,000 infants born per day and 152,000 people die per day, the net result is
    212,000 new inhabitants on Earth each day. What is the approximate net gain of new
    inhabitants on Earth in one year?
    a. 1 million
    b. 7 million
    c. 9.6 billion
    d. 10.5 billion
    e. 77 million

12. A hypothesis is
    a. testable.
    b. precise.
    c. an observation.
    d. also known as a scientific law.
    e. false.

Match the following words to their correct definition
_____13. Accuracy   a. how close to one another the repeated measurements are
_____14. Precision   b. how much a measured value differs from a true value
_____15. Uncertainty   .   c. how close a measured value is to the actual or true value

16. An activist group is protesting a landfill that is being built on a side of town with a low socio-
    economic status. The group states that by building the landfill there, it is violating the
    peoples' rights. This argument is from a field of study known as
    a. social justice.
    b. environmental racism.
    c. environmental justice.
    d. environmental subjectivity.
    e. environmental economics.

17. What type of chemical reaction occurs when one atom loses an electron and another atom gains the electron?
    a. Covalent bond
    b. Ionic bond
    c. Hydrogen bond
    d. Polar bond
    e. Energy bond

18. Household bleach with a pH of 13 is how many times more basic than pure water with a pH of 7?
    a. 100
    b. 1,000
    c. 6
    d. 6,000
    e. 1,000,000

19. Energy in the form of coal is an example of
    I. potential energy
    II. kinetic energy
    III. the first law of thermodynamics
        a. I only
        b. II only
        c. III only
        d. I and II
        e. I, II, and III

20. As you eat your lunch and you change the pizza into energy in the form of heat, you are demonstrating
    a. potential energy.
    b. kinetic energy.
    c. energy of motion.
    d. the first law of thermodynamics.
    e. the second law of thermodynamics.

21. If energy was not constrained under the second law of thermodynamics we would have what % of efficiency?
    a. 10
    b. 50
    c. 75
    d. 100
    e. 0

22. A sample of radioactive waste has a half life of 20 years and an activity level of 4 curies. After how many years will the activity level of this sample be 0.5 curies?
    a. 40 years
    b. 60 years
    c. 80 years
    d. 100 years
    e. 120 years

23. The type of system that tries to bring back homeostasis is a
    a. positive feedback loop.
    b. steady state.
    c. negative feedback loop.
    d. kinetic energy.
    e. potential energy.

24. According to the laws of thermodynamics
    I.   Energy is neither created nor destroyed.
    II. When energy is transformed, the quantity of energy remains the same.
    III. When energy is transformed, its ability to do work diminishes.
    a. I only
    b. II only
    c. III only
    d. I and III only
    e. I, II, and III

25. Most energy on Earth comes from
    a. plants.
    b. hydroelectric power.
    c. geothermal sources.
    d. the Sun.
    e. entropy.

Chapter 3: Ecosystem Ecology

Summary

This chapter is of critical importance on the AP exam. Make sure you are very familiar with the carbon, nitrogen, hydrologic and phosphorous cycles. There are usually at least five questions on the test that come directly from those cycles. Other important concepts are trophic pyramids, the formulas for photosynthesis and cellular respiration, and food webs and chains.

While You Read

Ecosystem ecology examines interactions between the living and nonliving world

Key Ideas

- All components of an ecosystem are interrelated.
- It is difficult to tell where one ecosystem ends and another begins.

Fill-in-the-Blank

1. An ecosystem is a particular _____ on Earth distinguished by its particular mix of interacting biotic and abiotic components.

Energy flows through ecosystems

Key Ideas

- Through photosynthesis, producers use solar energy to convert carbon dioxide and water into glucose; this process also produces oxygen as a waste product.
- There are producers, primary consumers, secondary consumers, tertiary consumers and decomposers in ecosystems.
- Ecological efficiency is only about 10%.

Match These Key Terms

_____1. Producers (autotrophs)

_____2. Photosynthesis

_____3. Cellular respiration

_____4. Consumers (heterotrophs)

_____5. Primary consumers

_____6. Secondary consumers

_____7. Tertiary consumers

_____8. Trophic levels

_____9. Food chain

_____10. Food web

a. Organisms that specializes in breaking down dead tissues and waste products into smaller particles

b. The total mass of all living matter in a specific area

c. Organisms that must obtain its energy by consuming other organisms

d. Organisms that use the energy of the Sun to produce usable forms of energy

e. A complex model of how energy and matter move between trophic levels

f. The process by which cells convert glucose and oxygen into energy, carbon dioxide, and water

g. The total amount of solar energy that producers in an ecosystem capture via photosynthesis over a given amount of time

h. Fungi or bacteria that recycle nutrients from dead tissues and wastes back into an ecosystem

i. Carnivores that eats secondary consumers

j. Carnivores that consumers dead animals

_____11. Scavengers

_____12. Detritivores

_____13. Decomposers

_____14. Gross primary productivity

_____15. Net primary productivity
_____16. Biomass

_____17. Standing crop

_____18. Ecological efficiency
_____19. Trophic pyramid

k. A representation of the distribution of biomass, numbers, or energy among trophic levels

l. The process by which producers use solar energy to convert carbon dioxide and water into glucose

m. Individuals incapable of photosynthesis; must obtain energy by consuming other organisms

n. The amount of biomass present in an ecosystem at a particular time

o. Carnivores that eats primary consumers

p. The sequence of consumption from producers through tertiary consumers

q. The proportion of consumed energy that can be passed from one trophic level to another

r. Levels in the feeding structure of organisms.

s. The energy captured by producers in an ecosystem minus the energy producers respire

Matter cycles through the biosphere
Key Ideas
- The major biogeochemical cycles are the hydrologic cycle, carbon cycle, nitrogen cycle, and the phosphorous cycle.

Match These Key Terms

_____1. Biosphere

_____2. Biogeochemical cycles

_____3. Hydrologic cycle

_____4. Transpiration

_____5. Evapotranspiration

_____6. Runoff

_____7. Macronutrients

_____8. Limiting nutrient

_____9. Nitrogen fixation
_____10. Leaching

a. The release of water from leaves during photosynthesis

b. A process by which some organisms can convert nitrogen gas molecules directly into ammonia

c. The transportation of dissolved molecules through the soil via groundwater

d. The region of our planet where life resides, the combination of all ecosystems on Earth

e. The six key elements that organisms need in relatively large amounts: nitrogen, phosphorous, potassium, calcium, magnesium, and sulfur

f. Water that moves across the land surface and into streams and rivers

g. The movements of matter within and between ecosystems

h. A nutrient required for the growth of an organism but available in a lower quantity than other nutrients

i. The movement of water through the biosphere

j. The combined amount of evaporation and transpiration

Ecosystems respond to disturbance
Key Ideas
- Ecosystem disturbances can be both natural and anthropogenic.
- How resilient an ecosystem is will determine how quickly it can recover from a disaster.

Match These Key Terms

_____1. Disturbance

_____2. Water shed

_____3. Resistance

_____4. Resilience

_____5. Restoration ecology

_____6. Intermediate disturbance hypothesis

a. The study and implementation of restoring damaged ecosystems

b. All land in a given landscape that drains into a particular stream, river, lake, or wetland

c. A measure of how much a disturbance can affect flows of energy and matter in an ecosystem

d. The hypothesis that ecosystems experiencing intermediate levels of disturbance are more diverse than those with high or low disturbance levels

e. An event, caused by physical, chemical, or biological agents, resulting in changes in population size or community composition

f. The rate at which an ecosystem returns to its original state after a disturbance

Ecosystem provide valuable services

Key Ideas

- Some valuable services that ecosystems provide are food, clean water filtering, lumber, medicine, oxygen production, and carbon sinks.

Match These Key Terms

_____1. Instrumental value
_____2. Intrinsic value

_____3. Provisions

a. Goods that humans can use directly

b. Something that has worth independent of any benefit it may provide to humans

c. Something that has worth as an instrument or a tool that can be used to accomplish a goal

Chapter Review

This chapter is all about the basics of environmental science. It is the foundation that is needed to grasp more difficult concepts you will be taught. Make sure you commit to memory the cycles that are covered. These will come up many times throughout the year and if you spend the time memorizing the steps and characteristics of each cycle now, you will save many hours of frustration later.

List questions from your initial reading of the chapter

_____

_____

_____

_____

_____

_____

_____

 **After You Read**

Short Answer

1. Write the formula for photosynthesis and describe the flow of energy from the Sun to living organisms.

2. Write the formula for cellular respiration.

3. Draw figure 3.9 to show the amount of energy that is present at each trophic level. How does the amount of energy at each level relate to the second law of thermodynamics?

4. Explain the difference between fast and slow carbon.

5. Briefly describe the steps of the nitrogen cycle. At which stage do bacteria play a role?

Do The Math

A farmer has decided to plant and sell Christmas trees. She estimates that each tree will sell for $75. How much will she make if she sells 300 trees?

If each tree take 2 gallons of water per day to thrive, how many gallons of water will it take to the grow the trees each year?

If a new sapling costs $1.25 and the farmer wants to replant her farm at the end of the year with 300 new trees, how much will the farmer need to invest?

Chapter 4: Global Climates and Biomes

Summary
This chapter explains the Earth's weather and climate. It covers the atmosphere, air and water currents, and the different terrestrial and aquatic biomes on the planet.

 **While You Read**

Global processes determine weather and climate
Key Ideas

- The processes that affect heat and precipitation around the globe are unequal heating of Earth by the Sun, atmospheric convection currents, the rotation of Earth, Earth's orbit around the Sun on a tilted axis, and ocean currents.
- The four layers of Earth's atmosphere are the troposphere, stratosphere, mesosphere, and thermosphere. The exosphere is space.
- The different types of ocean currents are gyres, upwellings, thermohaline circulation, heat transport, and the El Niño-southern oscillation.

Match These Key Terms

_____1. Climate

_____2. Troposphere

_____3. Stratosphere

_____4. Albedo

_____5. Saturation point

_____6. Adiabatic cooling

_____7. Adiabatic heating

_____8. Latent heat release

_____9. Hadley cells

_____10. Intertropical convergence zone

_____11. Polar cells

_____12. Coriolis effect

_____13. Gyres

a. The percentage of incoming sunlight reflected from a surface

b. Large scale patterns of water circulation that moves clockwise in the Northern Hemisphere and counterclockwise in the Southern Hemisphere

c. A region with dry conditions found on the leeward side of a mountain range as a result of humid winds from the ocean causing precipitation on the windward side.

d. The release of energy when water vapor in the atmosphere condenses into liquid water

e. Convection currents in the atmosphere that cycle between the equator and 30 degrees N and S

f. The average weather that occurs in a given region over a long period of time

g. The heating effect of increased pressure on air as it sinks toward the surface of Earth and decreases in volume

h. The cooling effect of reduced pressure on air as it rises higher in the atmosphere and expands

i. The deflection of an object's path due to the rotation of Earth

j. An oceanic circulation pattern that drives the mixing of surface water and deep water

k. A layer of the atmosphere closest to the surface of Earth, extending up to approximately 16 km and containing most of the atmosphere's nitrogen, oxygen, and water vapor

l. The upward movement of ocean water toward the surface as a result of diverging currents

m. An area of Earth that receives the most intense sunlight; where the ascending branches of the two

_____14. Upwelling

Hadley cells converge

n. The layer of the atmosphere above the troposphere, extending roughly 16 to 50 km above the surface of Earth

_____15. Thermohaline circulation

o. The periodic changes in winds and ocean currents, causing cooler and wetter conditions in the southeastern United States and unusually dry weather in southern Africa and Southeast Asia

_____16. El-Nino-Southern Oscillation

p. Convection cells in the atmosphere, formed by air that rises at 60 degrees N and S and sinks at the poles

_____17. Rain shadow

q. The maximum amount of water vapor that can be in the air at a given temperature

Variations in climate determine the dominant plant growth forms of terrestrial biomes

Key Ideas

- Biomes are categorized by particular combinations of average annual temperature and annual precipitation.
- The terrestrial biomes include tundra, boreal forest, temperate rainforest, temperate seasonal forest, woodland/shrubland, temperate grassland/cold desert, tropical rainforest, tropical seasonal forest/savanna, and subtropical desert.

Match These Key Terms

_____1. Biomes

a. A biome characterized by cold, harsh winters, and hot, dry summers

_____2. Tundra

b. A coastal biome typified by moderate temperatures and high precipitation

_____3. Permafrost

c. Geographic regions categorized by a particular combination of average annual temperature, annual precipitation, and distinctive plant growth forms on land

_____4. Boreal forest

d. A biome prevailing at approximately 30 degrees N and S, with hot temperatures, extremely dry conditions, and sparse vegetation

_____5. Temperate rainforest

e. A cold and treeless biome with low-growing vegetation

_____6. Temperate seasonal forest

f. A warm and wet biome found between 20 degrees N and S of the equator, with little seasonal temperature variation and high precipitation

_____7. Woodland/shrubland

g. An impermeable, permanently frozen layer of soil

_____8. Temperate grassland/cold desert

h. A biome marked by warm temperatures and distinct wet and dry seasons

_____9. Tropical rainforests

i. A forest made up primarily of coniferous evergreen trees that can tolerate cold winters and short growing seasons

_____10. Tropical seasonal forests/ savannas

j. A biome characterized by hot, dry summers and mild, rainy winters

_____11. Subtropical desert

k. A biome with warmer summers and colder winters than temperate rainforests and dominated by deciduous trees

Aquatic biomes are categorized by salinity, depth, and water flow

Key Ideas
- The aquatic biomes include steams/rivers, lakes/ponds, freshwater wetlands, salt marshes, mangrove swamps, intertidal zones, coral reefs and the open ocean.

Match These Key Terms

_____1. Littoral zone

_____2. Limnetic zone

_____3. Phytoplankton

_____4. Profundal zone

_____5. Benthic zone

_____6. Freshwater wetlands

_____7. Salt marshes

_____8. Mangrove swamps

_____9. Intertidal zone

_____10. Coral reefs

_____11. Coral bleaching

_____12. Photic zone

_____13. Aphotic zone

_____14. Chemosynthesis

a. The most diverse marine biomes on Earth, found in warm, shallow waters beyond the shoreline

b. A process used by some bacteria in the ocean to generate energy with methane and hydrogen sulfide

c. The shallow zone of soil and water in lakes and ponds where most algae and emergent plants grow

d. A zone of open water in lakes and ponds where rooted plants can no longer survive

e. The layer of ocean water that lacks sufficient sunlight for photosynthesis

f. Swamps that occur along tropical and subtropical coasts, and contain salt-tolerant trees with roots submerged in water

g. A region of water where sunlight does not reach, below the Limnetic zone in very deep lakes

h. A phenomenon in which algae inside corals die, causing the corals to turn white

i. Aquatic biomes that are submerged or saturated by water for at least part of each year, but shallow enough to support emergent vegetation

j. The upper layer of water in the ocean that receives enough sunlight for photosynthesis

k. Floating algae

l. The narrow band of coastline between the levels of high tide and low tide

m. Marshes containing nonwoody emergent vegetation, found along the coast in temperate climates

n. The muddy bottom of a lake or pond beneath the limnetic and profundal zones

Chapter Review

The most important concepts in this chapter for the AP exam are the layers of Earth's atmosphere, how temperature changes in each layer, the processes that affect climate, and how ocean currents influence weather and climate on Earth. You should be familiar with the different types of biomes (both aquatic and terrestrial), but do not need to memorize temperature and precipitation patterns or the animals and plants in each. As long as you have a general understanding of the biomes, you should be able to answer most questions that pertain to them.

List questions from your initial reading of the chapter

_____

_____

_____

_____

_____

After You Read

Short Answer

1. What are the five processes that determine climate?

2. Draw figure 4.1 below making sure to label the axis and include the temperature line.

3. What three phenomena cause wind patterns worldwide?

4. Summarize how the earth's oceans help regulate temperature on the planet.

5. What are the major terrestrial biomes on the Earth? Include a brief description of the general climate and plant growth for each biome.

6. What are the major aquatic biomes on the Earth?

Chapter 5: Evolution of Biodiversity

Summary
This chapter explains evolution and the processes that have created biodiversity on our planet.

 **While You Read**

Earth is home to a tremendous diversity of species

Key Ideas
- The three levels of biodiversity are ecosystem, species and genetic diversity.
- Species richness and species evenness are two different measures of species diversity.

Match These Key Terms

_____1. Ecosystem diversity
_____2. Species diversity
_____3. Genetic diversity
_____4. Species richness
_____5. Species evenness
_____6. Phylogeny

a. The number of species in a given area
b. The variety of ecosystems within a given region
c. The branching patterns of evolutionary relationships
d. The variety of species within a given ecosystem
e. The relative proportion of different species in a given area
f. The variety of genes within a given species

Evolution is the mechanism underlying biodiversity

Key Ideas
- The three ways evolution can occur is by artificial selection, natural selection, and random processes.
- Evolution by random processes can occur by mutations, genetic drift, the bottleneck effect, and the founder effect.

Match these Key Terms

_____1. Evolution

_____2. Microevolution

_____3. Macroevolution

_____4. Genes

_____5. Genotype

_____6. Mutation

_____7. Recombination

_____8. Phenotype

_____9. Evolution by artificial selection

_____10. Evolution by natural selection

a. A change in the genetic composition of a population over time as a result of humans selecting which individuals breed, typically with a preconceived set of traits in mind
b. Evolution that gives rise to new species, genera, families, classes, or phyla
c. A reduction in the genetic diversity of a population caused by a reduction in its size
d. Traits that improve an individual's fitness
e. Physical locations on the chromosomes within each cell of an organism
f. A change in a population descended from a small number of colonizing individuals
g. A random change in the genetic code produced by a mistake in the copying process
h. A change in the genetic composition of a population over time
i. A change in the genetic composition of a population over time as a result of the environment determining which individuals are most likely to survive and reproduce
j. Evolution occurring below the species level

_____11. Fitness

_____12. Adaptations

_____13. Genetic drift

_____14. Bottleneck effect

_____15. Founder effect

k. A change in the genetic composition of a population over time as a result of random mating

l. The complete set of genes in an individual

m. An individual's ability to survive and reproduce

n. A set of traits expressed by an individual

o. The genetic process by which one chromosome breaks off and attaches to another chromosome during reproductive cell division

Speciation and extinction determine biodiversity

Key Ideas

- When a species is geographically isolated from other populations, two distinct species can form.
- Natural evolution is an extremely slow process.
- Artificial evolution can be incredibly fast.

Match These Key Terms

_____1. Geographic isolation

_____2. Reproductive isolation

_____3. Allopatric speciation

_____4. Sympatric speciation

_____5. Genetic engineering

_____6. Genetically modified organisms

a. The result of two populations within a species evolving separately so they can no longer interbreed and produce viable offspring

b. When scientists copy genes from a species with desirable traits

c. An organism produced by copying genes from a species with a desirable trait and inserting them into another species

d. Physical separation of a group of individuals from others of the same species

e. The evolution of one species into two, without geographic isolation

f. The process of speciation that occurs with geographic isolation

Evolution shapes ecological niches and determines species distributions

Key Ideas

- Animals have a set of ideal conditions known as their fundamental niche.
- As environments change, species can adapt, move, or become extinct.
- Scientists use the fossil record to study species that lived millions of years ago.
- There have been five major times when the Earth has experienced mass extinctions.
- Scientists believe we are in the sixth mass extinction.

Match These Key Terms

_____1. Range of tolerance

_____2. Fundamental niche

_____3. Realized niche

_____4. Distribution

_____5. Niche generalist

_____6. Niche specialist

_____7. Fossils

_____8. Mass extinction

a. Areas of the world in which a species lives

b. The suite of ideal environmental conditions for a species

c. Remains of organisms that have been preserved in rock

d. The limits to the abiotic conditions that a species can tolerate

e. A large extinction of species in a relatively short period of time

f. The range of abiotic and biotic conditions under which a species actually lives

g. A species that can live under a wide range of abiotic or biotic conditions

h. A species that is specialized to live in a specific habitat or to feed on a small group of species

 **Chapter Review**

A few questions on evolution have appeared on past AP exams. For the exam, you mostly need to understand the big concepts. It is also important to understand the different levels of biodiversity (ecosystem, species and genetic) and the difference between species evenness and species richness. Also, know that populations can change into two genetically distinct populations by both natural means—such as a river—and also by human creations like cities.

List questions from your initial reading of the chapter

_____

_____

_____

_____

_____

_____

_____

_____

_____

_____

 **After You Read**

Short Answer

1. Explain in your own words the difference between species evenness and species richness.

2. What are the key ideas of Darwin's theory of evolution by natural selection?

3. List and describe the four random processes that can cause evolution.

4. Summarize in your own words the example of allopatric speciation from Figure 5.13.

5. What are the factors that determine the pace of evolution?

6. What five factors do scientists contend are causing the sixth mass extinction?

Review Practice Questions: Chapters 3-5

The Living World

1. Which of the following is the correct equation for photosynthesis?
    a. Energy + $6H_2O$ + $7 CO_2$ → $C_6H_{12}O_6$ + $8O_2$
    b. Energy + $6 H_2O$ + $6 CO_2$ → $C_6H_{12}O_6$ + $6 O_2$
    c. Solar energy + $6 H_2O$ + $8 CO_2$ → $C_6H_{12}O_6$ + $8 O_2$
    d. Solar energy + $8 H_2O$ + $8 CO_2$ → $C_6H_{12}O_6$+ $12 O_2$
    e. Solar energy + $6 H_2O$ + $6 CO_2$ → $C_6H_{12}O_6$ + $6 O_2$

2. Which of the following is the correct flow of energy in an ecosystem?
    a. Producer → Herbivore → Carnivore → Scavenger
    b. Producer → Scavenger → Herbivore → Carnivore
    c. Scavenger → Carnivore → Herbivore → Producer
    d. Carnivore → Scavenger → Herbivore → Producer
    e. Scavenger → Producer → Herbivore → Carnivore

3. The net primary productivity of an ecosystem is 25 kg C/m2/year, and the energy needed by the producers for their own respiration is 30 kg C/m2/year. The gross primary productivity of such an ecosystem would be
    a. 5 kg C/m2/year.
    b. 10 kg C/m2/year.
    c. 25 kg C/m2/year.
    d. 55 kg C/m2/year.
    e. 30 kg C/m2/year.

4. An ecosystem has an ecological efficiency of 10 percent. If the tertiary consumer level has 1 kcal of energy, how much energy did the producer level contain?
    a. 100 kcal
    b. 1000 kcal
    c. 10,000 kcal
    d. 90 kcal
    e. 900 kcal

5. Which of the following is NOT a part of the hydrologic cycle?
    a. Transpiration
    b. Condensation
    c. Solartranspiration
    d. Runoff
    e. Infiltration

Match the following processes with the correct product.

_____ 6. Nitrogen fixation
_____ 7. Nitrification
_____ 8. Assimilation
_____ 9. Ammonification
_____ 10. Denitrification

a. Nitrogen is assimilated into plant tissues
b. Nitrogen is released as a gas
c. Ammonia is converted into nitrite and nitrate
d. Bacteria convert ammonia into ammonium
e. Decomposers use waste as a food source and excrete ammonium

11. In which layer of Earth's atmosphere does the important ozone layer appear?
   a. Troposphere
   b. Stratosphere
   c. Mesosphere
   d. Thermosphere
   e. Exosphere

12. Average temperature and precipitation is known as
   a. rainfall.
   b. albedo.
   c. adiabatic cooling.
   d. adiabatic heating.
   e. climate.

13. The prevailing wind systems of the world are produced by
   a. convection currents and the Coriolis effect.
   b. convection currents and adiabatic cooling.
   c. convection currents and adiabatic heating.
   d. the Earth's rotation and the Coriolis effect.
   e. ocean circulation and the Coriolis effect.

14. In which month would the Sun be directly overhead at the equator?
   a. June
   b. March
   c. February
   d. October
   e. December

15. The El Niño—Southern Oscillation would bring what type of weather conditions to the southeastern United States?
   a. warmer, drier
   b. warmer, wetter
   c. cooler, drier
   d. cooler, wetter
   e. warmer, windier

Use the following climate diagram to answer questions 16-17.

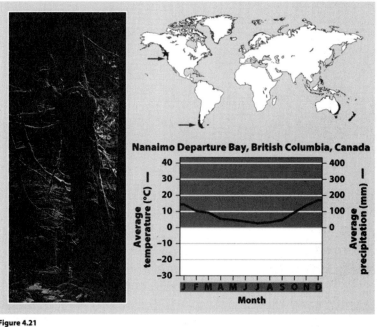

Figure 4.21
Environmental Science
© 2012 W. H. Freeman and Company

16. At which point on the graph is temperature at the highest and precipitation at the lowest?
    a. January
    b. April
    c. July
    d. October
    e. December

17. What month is average rainfall approximately 20 mm?
    a. January
    b. March
    c. July
    d. October
    e. December

18. Which of the following is NOT a measure of biodiversity?
    a. Ecosystem diversity
    b. Genetic diversity
    c. Species diversity
    d. Species richness
    e. Species wealth

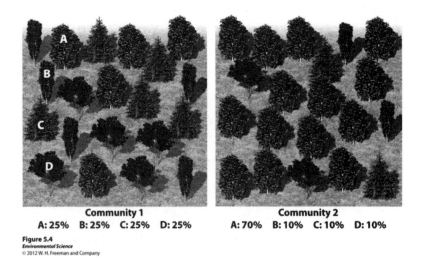

Community 1
A: 25%   B: 25%   C: 25%   D: 25%

Community 2
A: 70%   B: 10%   C: 10%   D: 10%

Figure 5.4
Environmental Science
© 2012 W. H. Freeman and Company

19. Which of the following is true of the picture above?
    a. Community 1 is less diverse than community 2
    b. Community 1 is more diverse than community 2
    c. Community 1 has a greater evenness and a greater richness than community 2
    d. Community 1 has a lower evenness and a lower richness than community 2
    e. Community 2 has a greater evenness but a lower richness than community 1

20. Which of the following is NOT a process by which species evolve?
    a. Mutation
    b. Genetic drift
    c. Bottleneck effect
    d. Genetic merging
    e. Founder effect

21. The type of speciation that could occur because of a mountain or other geographic isolation is known as
    a. reproductive isolation.
    b. allopatric speciation.
    c. sympatric speciation.
    d. bottleneck effect.
    e. founder effect.

22. A genetically modified organism is produced to
    I. defend plants against being eaten by insects.
    II. create mono-crops with higher yields.
    III. allow species to find new niches.
    a. I
    b. I and II
    c. I, II, and III
    d. I and III
    e. II and III

23. Which of the following would NOT be an example of a species niche?
    a. Evolution
    b. Competitors
    c. Habitat
    d. Food source
    e. Plant species

24. What type of rock would you most likely find a fossil?
    a. Igneous
    b. Sedimentary
    c. Metamorphic
    d. Mineral
    e. Marble

25. The current mass extinction is caused by all of the following EXCEPT
    a. habitat destruction.
    b. overharvesting.
    c. invasive species.
    d. genetic breeding.
    e. climate change.

Chapter 6: Population and Community Ecology

Summary
The principle objective of this chapter is to understand the dynamics behind population growth and the factors that drive growth.

While You Read

Nature exists at several levels of complexity
Key Ideas
- The levels of complexity from individual to biosphere are: individual, population, community, ecosystem, and biosphere.

Fill-in-the-Blanks
1. The simplest level is the _____, a single organism.
2. A population is composed of all individuals that belong to the same _____ and live in a given _____ at a particular _____.
3. A community incorporates all the _____ within a given _____.
4. Communities exist within an _____, which is all the biotic and abiotic components in a particular location.
5. The largest and most complex system environmental scientists study is the _____.

Population ecologists study the factors that regulate population abundance and distribution
Key Ideas
- Basic population characteristics include size, density, distribution, sex ratio, and age structure
- The three types of population distributions are random, uniform and clumped
- Density-dependent and density-independent factors can influence population size

Match These Key Terms

_____1. Population ecology

_____2. Population Size
_____3. Population density
_____4. Population distribution
_____5. Sex ratio
_____6. Age structure
_____7. Density-dependent factors

_____8. Limiting resource
_____9. Carrying capacity (K)

_____10. Density-independent factors

a. Factors that have the same effect on an individual's probability of survival and the amount of reproduction at any population size
b. The ratio of males to females
c. Factors that influence an individual's probability of survival and reproduction in a manner that depends on the size of the population
d. The study of factors that cause populations to increase or decrease
e. The limit of how many individuals in a population the food supply can sustain
f. A description of how many individuals fit into particular age categories
g. A description of how individuals are distributed with respect to one another
h. A resource that a population cannot live without and occurs in quantities lower than the population would require to increase in size
i. The number of individuals per unit area (or volume) at a given time
j. The number of individuals within a defined area at a given time

Growth models help ecologists understand population changes

Key Ideas

- The different growth models used to explain changes in population size are exponential and logistic.
- Some populations experience cycles of overshoots and die-offs that oscillate around the carrying capacity.
- Predators play an important role in limiting population growth.
- The two reproductive strategies an organism has are classified as K-selected or R-selected.

Match These Key Terms

_____1. Growth rate

_____2. Intrinsic growth rate

_____3. Exponential growth model

_____4. J-shaped curve

_____5. Logistic growth model
_____6. S-shaped curve

_____7. Overshoot

_____8. Die-off

_____9. K-selected species

_____10. R-selected species

_____11. Survivorship curves

_____12. Corridors

_____13. Metapopulation

a. A growth model that estimates a population's future size after a period of time, based on the intrinsic growth rate and the number of reproducing individuals currently in the population

b. Strips of natural habitat that connect separated populations

c. Groups of spatially distinct populations that are connected by occasional movements of individuals between them

d. The number of offspring an individual an individual can produce in a given time period, minus the deaths of the individual or any of its offspring during the same period

e. A rapid decline in a population due to death

f. Species that have a high intrinsic growth rate, which often leads to population overshoots and die-offs

g. The curve of the exponential growth model when graphed

h. The shape of the logistic growth model when graphed

i. Species with a low intrinsic growth rates that cause the population to increase slowly until it reaches carrying capacity

j. The maximum potential for growth of a population under ideal conditions with unlimited resources

k. Graphs that represent the distinct patterns of species survival as a function of age

l. A growth model that describes a population whose growth is initially exponential, but slows as the population approaches the carrying capacity of the environment

m. When a population becomes larger than the environment's carrying capacity

Community ecologists study species interactions

Key Ideas

- Species interact together in four ways—completion, predation, mutualism, and commensalism.
- Sometimes the loss of one species can have a major effect on the entire community.

Match These Key Terms

_____1. Community ecology

_____2. Competition

_____3. Competitive exclusion principle

_____4. Resource partitioning

_____5. Predation

_____6. True predators

_____7. Herbivores

_____8. Parasites

_____9. Pathogen

_____10. Parasitoids

_____11. Mutualism

_____12. Commensalism

_____13. Symbiotic

_____14. Keystone species

_____15. Predator-mediated competition

_____16. Ecosystem engineers

a. Species that are far more important in its their community than their relative abundance might suggest

b. The use of one species as a resource by another species

c. Keystone species that create or maintain habitat for other species

d. The study of interactions between species

e. Predators that live on or in the organism it consumes

f. A situation in which two species divide a resource, based on differences in their behavior or morphology

g. An interaction between species that increases the chances of survival or reproduction for both species

h. The struggle of individuals to obtain a limiting resource

i. Competition in which a predator is instrumental in reducing the abundance of a superior competitor, allowing inferior competitors to persist

j. The principle stating that two species competing for the same limiting resource cannot coexist

k. Predators that consume plants as prey

l. A relationship of two species that live in close association with each other

m. A relationship between species in which one species benefits and the other species is neither harmed nor helped

n. Illness-causing bacterium, virus, or parasite

o. Predators that typically kill its prey and consumes most of what it kills

p. Organisms that lay eggs inside other organisms

The composition of a community changes over time

Key Ideas

- As communities change over time, they undergo either primary or secondary succession.

Match These Key Terms

_____1. Ecological succession

_____2. Primary succession

_____3. Secondary succession

_____4. Pioneer species

a. Ecological succession occurring on surfaces that are initially devoid of soil

b. The succession of plant life that occurs in areas that have been disturbed but have not lost their soil

c. The replacement of one group of species by another group of species over time

d. A species that can colonize new areas rapidly

The species richness of a community is influenced by many factors

Key Ideas
- Latitude, time, habitat size, and distance from other communities help determine the number and types of species present in a biome

Fill in the blanks
1. The theory of island biogeography demonstrates the importance of both _____ _____ and _____ in determining species richness.

Chapter Review

Make sure you have a good grasp of this chapter. You need to know how to read graphs that show exponential growth, logistic growth, carrying capacity and predator-prey relationships. You also need to know what factors influence how populations change over time.

List questions from your initial reading of the chapter

_____

_____

_____

_____

_____

_____

_____

_____

_____

_____

_____

_____

**After You Read**

Short Answer

1. Diagram the different population distributions in the boxes below.

Random   **Uniform**   **Clumped**

2. Give an example of a density-dependent and a density-independent factor.

3. Draw a J-shaped growth curve and a logistic growth curve in the boxes below.

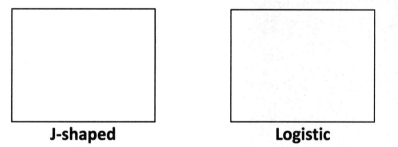

J-shaped   **Logistic**

4. Draw a typical carrying capacity graph where the population grows exponentially, overshoots the carrying capacity, has a die-off and then hovers around the carrying capacity.

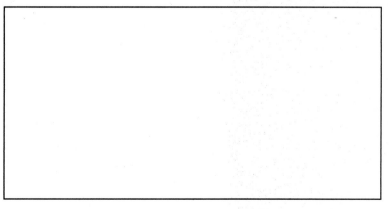

5. Summarize the predator-prey relationship in example 6.10 about the lynx and the hare.

6. Summarize the difference between mutualism, commensalism, and parasitism.

	WHO BENEFITS?	WHO IS HARMED?
Mutalism		
Commensalism		
Parasitism		

7. Explain the difference between primary succession and secondary succession in terms of cause and the types of plant species that begin to grow in each.

Do The Math

A population of deer has an initial population size of 15 individuals ($N_0=15$). Assume that the intrinsic rate of growth for a deer is r=.25 or 25%, which means that each deer produces a net increase of .25 deer each year.

With this information, predict the size of the deer population 1 year from now. (Show your work.) [Formula: $N_t = N_0 e^{rt}$]

How large will the deer population be 5 years from now?

How large will the deer population be 10 years from now?

Chapter 7: The Human Population

Summary
The objective of this chapter is to understand human population growth. This chapter includes the social, economic and environmental factors that determine growth. The amount of resource consumption of a growing human population is also addressed.

 **While You Read**

Scientists disagree on Earth's carrying capacity
Key Ideas
- The human population is growing exponentially.
- In 1798, Thomas Malthus said the human population was growing exponentially, while the food supply was growing linearly.

Many factors drive human population growth
Key Ideas
- Fertility rates, life expectancy, and migration rates all drive population growth.

Match These Terms

_____1. Demography
_____2. Demographers

_____3. Immigration
_____4. Emigration

_____5. Crude birth rate

_____6. Crude death rate
_____7. Doubling time
_____8. Total fertility rate
_____9. Replacement-level fertility

_____10. Developed countries
_____11. Developing countries
_____12. Life expectancy

_____13. Infant mortality rate
_____14. Child mortality rate
_____15. Age structure diagrams

_____16. Population pyramid

_____17. Population momentum

_____18. Net migration rate

a. Scientists in the field of demography
b. The difference between immigration and emigration in a given year per 1,000 people in a country
c. Countries with relatively high levels of industrialization and income
d. The movement of people into a country or region, having come from another country or region
e. A diagram that shows the numbers of individuals within each age category, typically expressed for males and females separately
f. The number of births per 1,000 individuals per year
g. The study of human populations and population trends
h. The number of deaths of children under age 5 per 1,000 live births
i. The movement of people out of a country or region, to settle in another country or region
j. Countries with relatively low levels of industrialization and income
k. The number of years it takes a population to double
l. Continued population growth that does not slow in response to growth reduction measures
m. Number of deaths per 1,000 individuals per year
n. Number of deaths of children under 1 year of age per 1,000 live births
o. An estimate of the average number of children that each woman in a population will bear throughout her childbearing years
p. An age structure diagram that is widest at the bottom and smallest at the top, typical of developing countries
q. The total fertility rate required to offset the average number of deaths in a population in order to maintain the current population size
r. The average number of years that an infant born in a particular year in a particular country can be expected to live, given the current average life span and death rate in that country

Many nations go through a demographic transition
Key Ideas
- The four phases of the demographic transition are phase 1 with slow population growth, phase 2 with rapid population growth, phase 3 with stable population growth and phase 4 with declining population growth.
- Some factors that can help to slow down population growth are education and affluence among females, family planning, and working women.

Fill-in-the-Blanks
1. The theory of demographic transition says that as a country moves from a _____

   _____ to _____ and increased _____,

   it undergoes a predictable shift in population growth.

2. Family planning is the regulation of the _____ or _____

   of offspring through the use of birth control.

Population size and consumption interact to influence the environment
Key Ideas
- Population size is a critical factor in the impact humans have on Earth.
- There are 6.8 billion people on Earth.
- Industrialized countries have a much greater resource use than non-industrialized countries.
- Countries with greater affluence, or wealth, have a much larger environmental income than poorer countries.

Match These Key Terms

_____1. Affluence

_____2. IPAT equation

_____3. Urban area

_____4. Gross domestic product

a. The state of having plentiful wealth; the possession of money, goods, or property

b. A measure of the value of all products and services produced in a country in a year

c. Impact = population X affluence X technology

d. An area that contains more than 385 people per square kilometer (1,000 people per square mile)

Sustainable development is a common, if elusive, goal
Key Ideas
- According to the Millennium Ecosystem Assessment, ecosystems will be threatened if the human population continues to grow, human well-being has improved for some people and declined for others, and if we establish sustainable practices, we may improve the standard of living for more people.

Chapter Review

This chapter is extremely important for the AP test. Make sure you can solve the population growth and IPAT equations and that you understand the steps of the demographic transition. Also, be familiar with the factors that can slow down population growth.

List questions from your initial reading of the chapter

_____

_____

_____

_____

_____

_____

_____

_____

_____

_____

 **After You Read**

1. Using the growth rate formula calculate the growth rate of a country with a CBR of 15, a CDR of 10, an immigration rate of 5, and an emigration rate of 2.

   Formula:
   growth rate = (Crude birth rate + Immigration) − (Crude death rate + emigration)
   $$\frac{}{10}$$

2. Calculate the doubling time of the country in problem #1 above.

   Formula: Doubling time (in years) = $\dfrac{70}{\text{Growth rate}}$

3. Draw the age structure diagrams below:

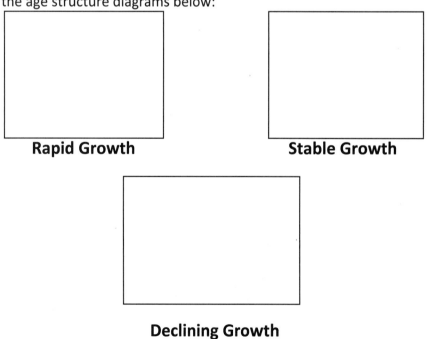

Rapid Growth   **Stable Growth**

Declining Growth

4. Sketch the demographic transition graph below.

Phase 1	Phase 2	Phase 3	Phase 4

5. Explain the IPAT equation.

6. What are the four types of economic activity that make up GDP?

Review Practice Questions: Chapters 6-7

Biological and Human Populations

1. All of the following are characteristics of populations EXCEPT
   a. population size can change depending on environmental factors.
   b. some populations remain relatively stable over time.
   c. population density can help determine management techniques.
   d. all populations are spread uniformly in respect to one another.
   e. populations are made up of individuals of varying ages.

2. If a population of 300 deer increases to 400 deer, the percent change is
   a. 3%.
   b. 33%.
   c. 300%.
   d. 50%.
   e. 25%.

3. Which of the following would be an example of a R-selected species?
   a. Humans
   b. Elephants
   c. Fish
   d. Dogs
   e. Oak trees

4. The maximum rate at which a population increases is known as its
   a. intrinsic growth rate.
   b. logistic growth rate.
   c. carrying capacity.
   d. linear growth rate.
   e. population oscillation.

5. All of the following are examples of predation EXCEPT
   a. a lion eating a gazelle.
   b. a clown fish living in a sea anemone.
   c. a gazelle eating grass.
   d. a tick living on a dog.
   e. larvae eating a host from the inside out.

6. A beaver that creates new ponds and wetland habitats would be known as a
   a. predator.
   b. indicator species.
   c. mutualistic species.
   d. successive species.
   e. keystone species.

7. Which of the following is true about ecological succession?
   a. Primary succession comes after secondary succession.
   b. Secondary succession is much slower than primary succession.
   c. During succession one species replaces another species over time.
   d. Secondary succession occurs when bare rock is colonized by organisms such as algae.
   e. A tornado would cause primary succession to occur.

8. Species richness would be determined by which of the following factors of island biogeography?
   I. island size
   II. how far away the island is from the mainland
   III. succession rates
   a. I and II
   b. II and III
   c. I and III
   d. I only
   e. I, II and III

9. Which of the following would be an example of resource partitioning?
   a. A wolf controlling populations of rabbits
   b. A tape worm living in the intestines of a dog
   c. Two bird species competing over a shared resources
   d. An orchid living on a tree branch
   e. Two bird species sharing resources by one eating large seeds and the other eating small seeds

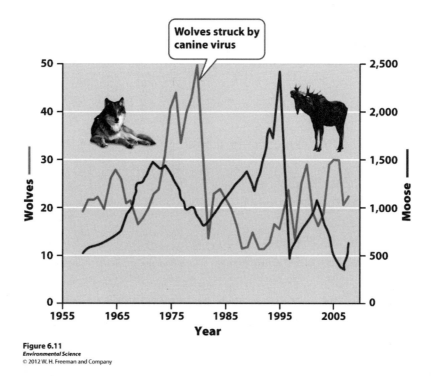

Figure 6.11
Environmental Science
© 2012 W. H. Freeman and Company

10. The graph above is an example of a
    a. mutualistic relationship.
    b. shared resource.
    c. K-selected species.
    d. predator-prey relationship.
    e. survivorship curve.

11. Thomas Malthus believed that
    a. nuclear war would end humankind.
    b. the human population size would exceed the food supply.
    c. technology would not be able to keep up with food supply.
    d. humans would not reach their carrying capacity due to technological advances.
    e. humans would reach their carrying capacity in the year 2100.

12. If a population of 10,000 has 300 births, 200 deaths, 50 immigrants and 60 emigrants, what is the populating growth rate?
    a. 0.9%
    b. 9%
    c. 90%
    d. 2.4%
    e. 24%

13. If a countries population growth rate is 5%, what is the countries doubling time?
    a. 5 years
    b. 35 years
    c. 14 years
    d. 42 years
    e. 72 years

14. Which of the following is the most critical factor in controlling the size of the human population?
    a. Lowering the total fertility rate
    b. Increasing life expectancy
    c. Providing for elderly care
    d. Increasing the overall health of humans
    e. Decreasing child/infant mortality

15. To determine how many people will be added to a population you would
    a. subtract births from deaths.
    b. multiply births and deaths.
    c. add all age categories and multiply by 1.2%.
    d. take the global population size and subtract the infant mortality rate.
    e. multiply the size of the population by the population's growth rate.

16. What are the two reasons for the rapid growth of the human population over the past 8,000 years?
    a. Lower infant and child mortality
    b. Increased medical and technological advances
    c. Food deficits and technology
    d. Advent of agriculture and The Plague
    e. Lack of reliable birth control and education

Use the following graphs to answer questions 17 & 18.

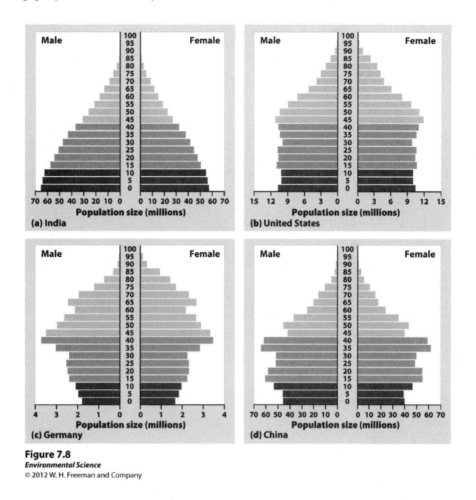

Figure 7.8
Environmental Science
© 2012 W. H. Freeman and Company

17. Which country's age structure diagram represents a population that was growing exponentially, but has since leveled off and is now decreasing in size?
    a. India
    b. United States
    c. Germany
    d. China
    e. none of the above

18. What can you tell about India's population from looking at the age structure diagram above?
    a. It is growing rapidly.
    b. It is industrialized.
    c. Its population momentum is slowing down.
    d. It has a large immigration rate.
    e. It has advanced medical care.

Use the following graph to answer questions 19 & 20.

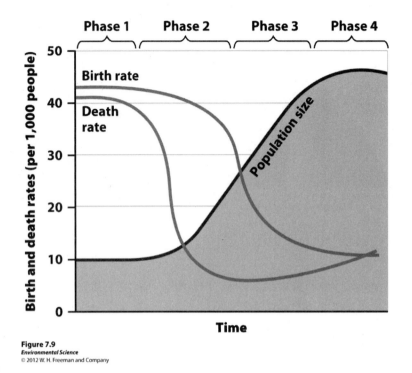

Figure 7.9
Environmental Science
© 2012 W. H. Freeman and Company

19. At which phase does the population size remain stable?
    a. Phase 1
    b. Phase 1 and 2
    c. Phase 3
    d. Phase 4
    e. Phase 1 & 4

20. At which phase is the population growing exponentially?
    a. Phase 1
    b. Phase 2
    c. Phase 3
    d. Phase 4
    e. Phase 1 and 2

21. If a developing nation quickly reduces its growth rate to 0%, its population would
    a. decrease rapidly.
    b. decrease slowly.
    c. level off.
    d. continue growing for many years then level off.
    e. grow exponentially.

22. A country with a large population that is relatively affluent and is technologically advanced will
    a. remain stable.
    b. need medical services.
    c. have a large environmental impact.
    d. have a low GDP.
    e. have a low emigration rate.

23. The agricultural revolution led to an
    a. increase in the size of the human population.
    b. increase in biological warfare.
    c. increase in infectious diseases.
    d. increase in noncommunicable diseases.
    e. increase in the baby boom.

24. At present the size of the earth's human population is nearing
    a. 300 million.
    b. 1 billion.
    c. 3.5 billion.
    d. 7 billion.
    e. 300 billion.

25. Gross domestic product is made up of all of the following factors EXCEPT
    a. consumer spending.
    b. population growth.
    c. investments.
    d. government spending.
    e. exports minus imports.

Chapter 8: Earth Systems

Summary

This chapter is all about earth science. It discusses Earth's geologic cycle—the tectonic cycle, the rock cycle, and the formation of soil.

While You Read

The availability of Earth's resources was determined when the planet formed

Key Ideas

- The Earth formed approximately 4.6 billion years ago.
- The elements and minerals that were present when the planet formed are all that we have.

Match These Key Terms

_____1. Core

_____2. Mantle

_____3. Magma

_____4. Asthenosphere

_____5. Lithosphere

_____6. Crust

a. The layer of Earth located in the outer part of the mantle, composed of semi-molten rock

b. In reference to Earth, the innermost layer

c. In geology, the chemically distinct outermost layer of the lithosphere

d. The outermost layer of Earth, including the mantle and crust

e. Molten rock

f. The layer of Earth above the core, containing magma

Earth is dynamic and constantly changing

Key Ideas

- The theory of plate tectonics states that the Earth's lithosphere is constantly moving.
- The plates move apart, together, or slide past each other.
- Volcanoes and earthquakes occur at plate boundaries.

Match These Key Terms

_____1. Hot spots

_____2. Plate tectonics

_____3. Tectonic cycle

_____4. Subduction

_____5. Volcano

_____6. Divergent plate boundaries

_____7. Seafloor spreading

_____8. Convergent plate boundaries

a. A vent in the surface of Earth that emits ash, gases, or molten lava

b. An area where tectonic plates move sideways past each other

c. The formation of new ocean crust as a result of magma pushing upward and outward from Earth's mantle to the surface

d. The cycle of processes that build up and break down the lithosphere

e. The theory that the lithosphere of Earth is divided into plates, most of which are in constant motion

f. The frequency and intensity of earthquakes

g. A scale that measures the largest ground movement that occurs during an earthquake

h. The exact point on the surface of Earth directly above the location where rock ruptures during an earthquake

_____9. Transform fault boundary

i. The sudden movement of Earth's crust caused by a release of potential energy along a geologic fault and usually causing a vibration or trembling at Earth's surface

_____10. Fault

j. A large expanse of rock where a fault has occurred

_____11. Fault zone

k. A fracture in rock caused by a movement of Earth's crust

_____12. Earthquakes

l. The process of one crustal plate passing under another

_____13. Seismic activity

m. In geology, places where molten material from Earth's mantle reaches the lithosphere

_____14. Epicenter

n. Areas beneath oceans where tectonic plates move away from each other

_____15. Richter scale

o. Areas where plates move toward one another and collide

The rock cycle recycles scarce minerals and elements
Key Ideas
- The rock cycle is the formation and destruction of rock.
- The three types of rocks are igneous, sedimentary, and metamorphic.

Match These Key Terms

_____1. Rock cycle

a. The mechanical breakdown of rocks and minerals

_____2. Minerals

b. The continuous formation and destruction of rock on and below the surface of Earth

_____3. Intrusive igneous rocks

c. Rocks that form when sediments such as muds, sands, or gravels are compressed by overlying sediments

_____4. Extrusive igneous rocks

d. Igneous rock that forms when magma rises up and cools in place underground

_____5. Fractures

e. Rock that forms when magma cools above the surface of Earth

_____6. Sedimentary rocks
_____7. Metamorphic rocks

f. In geology, cracks that occur in rock as it cools

g. Solid chemical substances with a uniform, often crystalline, structure that forms under specific temperatures and pressures

_____8. Physical weathering

h. The accumulation or depositing of eroded material such as sediment, rock fragments, or soil

_____9. Chemical weathering

i. Rocks that form when sedimentary rock, igneous rock, or other metamorphic rock are subjected to high temperature and pressure

_____10. Acid precipitation (acid rain)

j. The breakdown of rocks and minerals by chemical reactions, the dissolving of chemical elements from rocks, or both

_____11. Erosion
_____12. Deposition

k. Precipitation high in sulfuric acid and nitric acid from reactions between sulfur dioxide and water vapor and nitrogen oxides and water vapor in the atmosphere

l. The physical removal of rock fragments from a landscape or ecosystem

Soils link the rock cycle and the biosphere

Key Ideas

- Soil is important as a place for plant growth, as a habitat for other organisms, and as a recycling system for organic waste.
- Soils help filter and purify water.
- Soil can form from the breakdown of rock and the decomposition of organic matter.
- Sand, silt, and clay determine soil texture.
- Overuse of land, forestry, and other human activities are degrading soil.

Match These Key Terms

_____ 1. Soil

_____ 2. Parent material

_____ 3. Horizons

_____ 4. O horizon

_____ 5. A horizon (topsoil)

_____ 6. E horizon

_____ 7. B horizon (subsoil)

_____ 8. C horizon

_____ 9. Texture

_____ 10. Cation exchange capacity (CEC)

_____ 11. Base saturation

_____ 12. Soil degradation

a. The proportion of soil bases to soil acids, expressed as a percentage

b. Frequently the second major soil horizon, composed primarily of mineral material with very little organic matter

c. Frequently the top layer of soil, characterized by mixing of organic material and mineral material

d. The least-weathered soil horizon, which always occurs beneath the B horizon and is similar to the parent material

e. The ability of a particular soil to absorb and release cations

f. The zone of leaching that forms under the O horizon or, less often, the A horizon

g. Layers of soil

h. Rock underlying soil; the material from which the inorganic components of a soil are derived

i. A mix of geologic and organic components that form a dynamic membrane covering much of Earth's surface

j. The loss of some or all of a soil's ability to support plant growth

k. The property of soil determined by relative proportions of sand, silt, and clay

l. The organic horizon at the surface of many soils, composed of organic detritus in various stages of decomposition

The uneven distribution of mineral resources has social and environmental consequences

Key Ideas

- Silicon and oxygen are the two most abundant elements in the Earth's crust.
- There are two types of mining: surface and subsurface.

Match These Key Terms

_____1. Crustal abundance

_____2. Ores

_____3. Metals

_____4. Reserve

_____5. Strip mining
_____6. Mining spoils (tailings)

_____7. Open-pit mining

_____8. Mountaintop removal
_____9. Placer mining

_____10. Subsurface mining

a. In resource management, the known quantity of a resource that can be economically recovered

b. Concentrated accumulations of minerals from which economically valuable materials can be extracted

c. Mining techniques used when the desired resource is more than 100 m below the surface of Earth

d. A mining technique in which metals and precious stones are sought in river sediments

e. The average concentration of an element in Earth's crust

f. Elements with properties that all allow them to conduct electricity and heat energy, and perform other important functions

g. A mining technique that uses a large pit or hole in the ground, visible from the surface of Earth

h. The removal of strips of soil and rock to expose ore

i. A mining technique in which the entire top of a mountain is removed with explosives

j. Unwanted waste material created during mining

Chapter Review

Make sure that you can identify and understand the rock cycle, the layers of soil, and the environmental impacts of each type of mining. These topics show up each year on the AP Exam.

List questions from your initial reading of the chapter

_____

_____

_____

_____

_____

_____

_____

_____

_____

 **After You Read**

Short Answer

1. Draw a diagram of the Earth's layers. Use Figure 8.2a to help.

2. Explain the theory of plate tectonics.

3. Draw a diagram of each of the following:

Divergent Plate Boundary

Convergent Plate Boundary

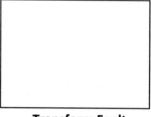

Transform Fault

4. Each time you move up the Richter scale you are increasing the damage done by an earthquake by a multiple of _____.

5. Explain the processes that form sedimentary rocks.

6. Diagram the layers of soil in the space below.

7. Name and describe the 5 factors that determine soil properties.

8. List three chemical properties and 3 physical properties of soil.

9. Explain in your own words what figure 8.23 is demonstrating.

10. Describe the differences between surface and subsurface mining.

Do The Math

Freemont is approximately 500 km north of Shaky Acres. If the plate under Shaky Acres is moving at 15mm per year towards Freemont, how long will it take for Shaky Acres to be located next to Freemont? [**Formula: time= distance/rate**]

Chapter 9: Water Resources

Summary
The main objectives of this chapter are to identify and describe water resources and to understand how humans are managing and using this valuable resource.

 **While You Read**

Water is abundant, but usable water is rare
Key Ideas
- 97% of water on Earth is saltwater.
- Freshwater (not in the form of ice and glaciers) is found in groundwater, surface water (rivers, lakes, ponds), and the atmosphere.

Match These Key Terms

_____1. Aquifers
    a. The land adjacent to a river

_____2. Unconfined aquifers
    b. Pavement or buildings that do not allow water penetration

_____3. Confined aquifers
    c. A process by which water percolates through the soil and works its way into an aquifer

_____4. Water table
    d. When a body of water becomes rich in nutrients

_____5. Groundwater recharge
    e. An aquifer made of porous rock covered by soil, which water can easily flow into and out of

_____6. Springs
    f. Wells created by drilling a hole into a confined aquifer

_____7. Artesian wells
    g. Permeable layers of rock and sediment that contain groundwater

_____8. Cone of depression
    h. When a body of water has moderate levels of productivity

_____9. Saltwater intrusion
    i. When a body of water has a low level of productivity as a result of low amounts of nutrients in the water

_____10. Floodplain
    j. The uppermost level at which the water in a given area fully saturates rock or soil

_____11. Oligotrophic
    k. An area from which the groundwater has been rapidly withdrawn

_____12. Mesotrophic
    l. An infiltration of salt water in an area where groundwater pressure has been reduced from extensive drilling of wells

_____13. Eutrophic
    m. An aquifer surrounded by a layer of impermeable rock or clay that impedes water flow

_____14. Impermeable surfaces
    n. Natural sources of water formed when water from an aquifer percolates up to the ground surface

Humans can alter the availability of water
Key Ideas
- Humans have altered the flow of rivers with levees, blocked the flow with dams, diverted water from rivers and lakes and removed salt from saltwater to make fresh water

Match These Key Terms

_____1. Levee

_____2. Dikes

_____3. Dam

_____4. Reservoir

_____5. Fish ladders

_____6. Aqueducts

_____7. Desalination

a. Canals or ditches used to carry water from one location to another

b. A barrier that runs across a river or stream to control the flow of water

c. Structures built to prevent ocean waters from flooding adjacent land

d. The process of removing the salt from salt water

e. Stair-like structures that allow migrating fish to get around a dam

f. An enlarged bank built up on each side of a river to prevent flooding

g. A body of water created by blocking the natural flow of a waterway

Water is used by humans for agriculture, industry, and household needs

Key Ideas

- 70 % of the world's freshwater is used to irrigate crops
- Agriculture, industry and households are the major water users

Fill-in-the-Blanks

Hydroponic agriculture is the cultivation of crop plants under greenhouse conditions with their _____ immersed in a _____ solution, but no _____.

The future of water availability depends on many factors

Key Ideas

- Water experts predict that as populations grow, conflicts over water will increase.
- Water conservation is needed in areas where water is rare.
- Examples of water conservation methods are low flow toilets, planting crops that do not need as much water, more efficient washing machines, more efficient manufacturing.

 # Chapter Review

The information in almost every section of this chapter has appeared on previous AP exams. Pay special attention to the three biggest uses of water: agricultural, industrial and municipal. Know what _eutrophic_ is and make sure you understand the effects nitrogen and phosphorous have on a water resource like a lake.

List questions from your initial reading of the chapter

_____

_____

_____

_____

_____

_____

 **After You Read**

Short Answer

1. Explain the difference between a confined and an unconfined aquifer.

2. What is saltwater intrusion and what causes this problem?

3. Describe the causes and effects of a eutrophic body of water.

4. Describe the ecological benefits of freshwater wetlands.

5. Describe some human activities that have impacted flooding.

6. Describe both the benefits and consequences of dams.

7. What are the three major users of water and what percentage of water is used for each?

Do The Math

You have just moved into a new house and need to purchase a washing machine. You've heard that front-loading machines use less water but that they are more expensive. A traditional machine costs $500 and a new front-loading washing machine costs $1,000.

If you wash 8 loads of laundry each week, how many loads do you wash each month (assume 4 weeks/month)?

If the traditional machine uses 200L of water per load and the new machine uses 100 L of water per load, how many liters of water would the new washing machine save per month?

If water costs $0.35 for every 1000L, how much money would be saved each month by using the new washing machine?

Review Practice Questions: Chapters 8-9

Earth Systems and Resources

1. The Earth's plates are in constant motion because
   a. the core is made up of iron.
   b. of convection currents found in the asthenosphere.
   c. the lithosphere being so dense.
   d. the core being so dense.
   e. subduction zones found near Japan and Argentina.

2. The result of Seafloor spreading is associated with which of the following plate boundaries
   a. divergent plate boundary.
   b. convergent plate boundary.
   c. transform fault boundary.
   d. subduction zone.
   e. hotspot.

3. Measured on the Richter scale, an earthquake with a magnitude of 6.0 is _____ times greater than an earthquake with a magnitude of 3.0.
   a. 10
   b. 100
   c. 1,000
   d. 10,000
   e. 100,000

4. Volcanoes might be found near a plate boundary that is
   a. diverging.
   b. converging.
   c. spreading.
   d. sliding next to one another.
   e. magnetic.

5. Plate movement occurs in which part of the earth's crust?
   a. Core
   b. Upper mantle
   c. Asthenosphere
   d. Lithosphere
   e. Lower mantle

6. Earthquakes can lead to
   I. collapsed structures and buildings
   II. contaminated water supplies
   III. loss of life
      a. I only
      b. II only
      c. III only
      d. I and III only
      e. I, II, and III

7. All of the following are ways sedimentary rock is formed EXCEPT
      a. weathering.
      b. erosion.
      c. transportation.
      d. compression.
      e. melting.

8. Fossils are most likely found in
      a. igneous rock.
      b. metamorphic rock.
      c. sedimentary rock.
      d. basalt.
      e. granite.

9. Acid precipitation with a pH of 6 can damage all of the following EXCEPT
      a. automotive coatings.
      b. concrete.
      c. limestone.
      d. marble.
      e. granite.

10. Some countries that do not have large quantities of fresh water get their water from
      a. icebergs.
      b. glaciers.
      c. dams.
      d. center-pivot irrigation.
      e. desalination.

11. Soil found in tropical rain forests are generally
      a. rich in organic material.
      b. rich in quartz sand.
      c. deep and porous.
      d. nutrient poor.
      e. acidic.

12. The soil layer most similar to the parent material is the
    a. O horizon.
    b. A horizon.
    c. B horizon.
    d. C horizon.
    e. E horizon.

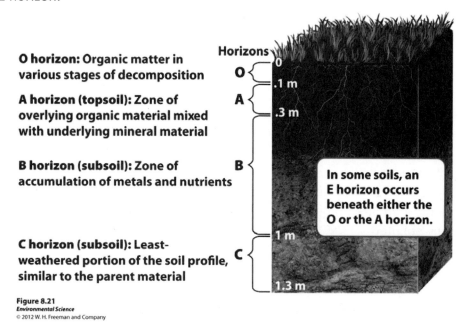

O horizon: Organic matter in various stages of decomposition

A horizon (topsoil): Zone of overlying organic material mixed with underlying mineral material

B horizon (subsoil): Zone of accumulation of metals and nutrients

C horizon (subsoil): Least-weathered portion of the soil profile, similar to the parent material

In some soils, an E horizon occurs beneath either the O or the A horizon.

Figure 8.21
Environmental Science
© 2012 W. H. Freeman and Company

13. The percentage of sand, silt, and clay determine the soils
    a. porosity.
    b. texture.
    c. chemical makeup.
    d. temperature.
    e. density.

14. Soil degradation is caused by all of the following EXCEPT
    a. machines.
    b. humans.
    c. livestock.
    d. plowing.
    e. compost.

15. The Surface Mining Control and Reclamation Act of 1977 regulates
    a. mining companies that harm their workers.
    b. land be minimally disturbed during the mining process.
    c. how much water a mine can use.
    d. the type of fill material that must be used.
    e. the equipment used while mining.

16. Ice and glaciers represent what percentage of the freshwater on the earth?
    a. 22%
    b. 0.5%
    c. 97%
    d. 3 %
    e. 77%

17. Confined aquifers
    a. are polluted more easily than unconfined aquifers.
    b. are recharged from rainwater.
    c. are impermeable.
    d. are typically found near the surface of the earth.
    e. are made of porous rock covered by soil.

18. The Ogallala aquifer is
    a. the largest aquifer in the United States.
    b. the largest aquifer in the world.
    c. heavily polluted.
    d. below the water table.
    e. contaminated by saltwater intrusion.

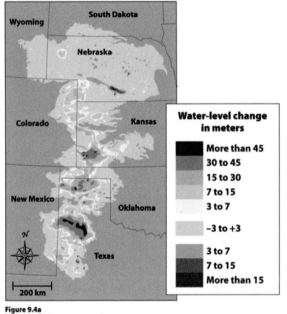

Figure 9.4a
Environmental Science
© 2012 W. H. Freeman and Company

19. The Nile River was used by the Egyptians because it
    a. was fast moving.
    b. was deep for dredging.
    c. provided clean water for drinking.
    d. made a fertile floodplain for agriculture.
    e. had large fish for a food source.

20. A lake that has been contaminated with high levels of nitrogen and phosphorous is
    a. oligotrophic.
    b. eutrophic.
    c. mesotrophic.
    d. impermeable.
    e. easily flooded.

21. Fish ladders were built to
    a. help fish get over a dam.
    b. breed fish in captivity.
    c. bring fish to undeveloped nations.
    d. stop invasive species from reproducing in public lakes.
    e. help fish reproduce in harsh environmental conditions.

22. Over-pumping of an aquifer can cause all of the following EXCEPT
    a. saltwater intrusion.
    b. cone of depression.
    c. eutrophication.
    d. dry wells.
    e. land and property loss.

23. Desalinization is
    a. used extensively across Europe.
    b. causing drought.
    c. a problem with saltwater intrusion.
    d. occurring at a rapid rate due to over irrigating farmland.
    e. helping water-poor countries obtain fresh water.

24. The type of irrigation technique that is over 95% efficient is
    a. spray irrigation.
    b. flood irrigation.
    c. furrow irrigation.
    d. drip irrigation.
    e. hydroponic irrigation.

25. The largest percentage of indoor household water use is from
    a. flushing toilets.
    b. bathing.
    c. laundry.
    d. cooking.
    e. drinking.

Chapter 10: Land, Public and Private

Summary
The key concepts of this chapter are the tragedy of the commons, public land management, and urban development.

While You Read

Human land use affects the environment in many ways

Key Ideas

- When something that is not owned is overexploited (overgrazed, overharvested, and deforested) for individual gain at the expense of the community, it is called the "tragedy of the commons."
- An externality is the cost or benefit of a good or service that is not included in the price of the item.
- We need to know a resource's maximum amount that can be harvested without harming the resource.

Match These Key Terms

_____1. Tragedy of the commons

_____2. Externality

_____3. Maximum sustainable yield

a. The cost or benefit of a good or service that is not included in the purchase price of that good or service

b. The maximum amount of a renewable resource that can be harvested without compromising the future availability of that resource

c. The tendency of a shared, limited resource to become depleted because people act from self-interest for short-term gain

Public lands are classified according to their use

Key Ideas

- The six categories of public land are national parks, managed resource protected areas, habitat/species management areas, strict nature preserves and wilderness areas, protected landscapes and seascapes, and national monuments.
- 42% of land in the United States is publicly owned.

Fill-in-the-Blanks

1. The resource conservation ethic states that people should _____ resource use

   based on the _____ good for everyone.

2. Some public lands are in fact classified as multiple-use lands, and may be used for _____,

   _____, _____ _____, and _____

   _____.

Land management practices vary according to land use
Key Ideas
- Overgrazing of rangelands can leave the land polluted and exposed to erosion, and make it difficult for soils to absorb and retain water when it rains.
- Clear-cutting and selective cutting are the two most common ways trees are harvested.
- Fire is a natural process that is important for nutrient cycling and regeneration.

Match These Key Terms

_____1. Rangelands

_____2. Forests

_____3. Clear-cutting

_____4. Selective cutting

_____5. Ecologically sustainable forestry

_____6. Tree plantations

_____7. Prescribed burn

_____8. National wildlife refuges

_____9. National wilderness areas

_____10. National Environmental Policy Act

_____11. Environmental impact statement

_____12. Environmental mitigation plan

_____13. Endangered Species Act

a. An approach to removing trees from forests in ways that do not unduly affect the viability of other trees

b. Land dominated by trees and other woody vegetation and sometimes used for commercial logging

c. A plan that outlines how a developer will address concerns raised by a project's impact on the environment

d. Federal public lands managed for the primary purpose of protecting wildlife

e. A document outlining the scope and purpose of a development project, describing the environmental context, suggesting alternative approaches to the project, and analyzing the environmental impact of each alternative

f. A 1973 U.S. act designed to protect species from extinction

g. Dry, open grasslands

h. The method of harvesting trees that involves the removal of single trees or a relatively small number of trees from among many in a forest

i. A fire deliberately set under controlled conditions in order to reduce the accumulation of dead biomass on a forest floor.

j. Large areas typically planted with a single rapidly growing tree species

k. Areas set aside with the intent of preserving a large tract of intact ecosystem or a landscape

l. A 1969 U.S. federal act that mandates an environmental assessment of all projects involving federal money or federal permits

m. A method of harvesting trees that involves removing all or almost all of the trees within an area

Residential land use is expanding
Key Ideas
- In the United States, the greatest percentage of population growth has occurred in suburban and exurban communities.
- The four main causes of urban sprawl in the United States are automobiles and highway construction, living costs, urban blight, and government policies.

Match These Key Terms

_____ 1. Suburban

_____ 2. Exurban

_____ 3. Urban sprawl

_____ 4. Urban blight

_____ 5. Highway Trust Fund

_____ 6. Induced demand
_____ 7. Zoning

_____ 8. Multi-use zoning

_____ 9. Smart growth

_____ 10. Stakeholders

_____ 11. Sense of place

_____ 12. Transit-oriented development

_____ 13. Infill

_____ 14. Urban growth boundaries

a. An area similar to a suburb, but unconnected to any central city or densely populated area

b. Development that fills in vacant lots within existing communities

c. The phenomenon in which increase in the supply of a good causes demand to grow

d. A zoning classification that allows retail and high-density residential development to coexist in the same area

e. The degradation of the built and social environments of the city that often accompanies and accelerates migration to the suburbs

f. Restrictions on development outside a designated area

g. Urbanized areas that spread into rural areas, removing clear boundaries between the two

h. People or organizations with an interest in a particular place or issue

i. An area surrounding a metropolitan center, with a comparatively low population density

j. The feeling that an area has distinct and meaningful character

k. A set of principles for community planning that focuses on strategies to encourage the development of sustainable, healthy communities

l. Development that attempts to focus dense residential and retail development around stops for public transportation, a component of smart growth

m. A planning tool used to separate industry and business from residential neighborhoods

n. A U.S. federal fund that pays for the construction and maintenance of roads and highways

Chapter Review

The concepts in this chapter that have shown up frequently on previous exams are the tragedy of the commons, negative externalities, and timber harvest practices. FRQs based on urban blight and on forest management and fires have also appeared on the test in the past.

List questions from your initial reading of the chapter

_____

_____

_____

_____

_____

_____

**After You Read**

Short Answer

1. What is the single largest cause of species extinctions today?

2. Give an example of a negative externality.

3. What are some of the uses of public and private land in the United States?

4. Describe the cons of fire suppression.

5. Describe urban sprawl and urban blight, including causes and environmental consequences of both.

Chapter 11: Feeding the World

Summary
This chapter focuses on the growing human population and the amount of food needed to sustain it. The objectives of this chapter are to understand hunger and malnutrition, agriculture, and the environmental impacts of different farming techniques.

While You Read
Human nutritional requirements are not always satisfied

Key Ideas
- Humans need approximately 2,200 kilocalories per day.
- Too many calories or not enough calories can lead to malnutrition.
- The primary reason for undernutrition and malnutrition is poverty.

Match These Key Terms

_____1. Undernutrition

_____2. Malnourished

_____3. Food security

_____4. Food insecurity

_____5. Famine

_____6. Anemia

_____7. Overnutrition

_____8. Meat

a. The condition in which food insecurity is so extreme that large numbers of deaths occur in a given area over a relatively short period

b. Ingestion of too many calories and improper foods

c. A condition in which people have access to sufficient, safe, and nutritious food that meets their dietary needs for an active and healthy life

d. A condition in which people do not have adequate access to food

e. Livestock or poultry consumed as food

f. A deficiency of iron

g. The condition in which not enough calories are ingested to maintain health

h. Having a diet that lacks the correct balance of proteins, carbohydrates, vitamins, and minerals

The Green Revolution and industrial farming methods have transformed agriculture

Key Ideas
- The abundance of food supplied by agriculture is one factor that has lead to the exponential growth of the human population.
- A great deal of energy goes into growing, harvesting, processing, and preparing food.
- The Green Revolution involves practices such as mechanization, irrigation, monocropping, and the use of fertilizers and pesticides.

Match These Key Terms

_____1. Industrial agriculture (agribusiness)

_____2. Energy subsidy

_____3. Green Revolution
_____4. Economies of scale

_____5. Waterlogging

_____6. Salinization

_____7. Organic fertilizers

_____8. Synthetic (inorganic) fertilizers

_____9. Monocropping
_____10. Pesticides

_____11. Insecticides

_____12. Herbicides
_____13. Broad-spectrum pesticides

_____14. Selective pesticides

_____15. Persistent pesticides

_____16. Bioaccumulation

_____17. Nonpersistent pesticide

_____18. Resistant

_____19. Pesticide treadmill

a. Pesticides that target species of insects and other invertebrates

b. An increased concentration of a chemical within an organism over time

c. Pesticides that kill many different types of pests

d. A shift in agricultural practices in the 20th century that included new management techniques, mechanization, fertilization, irrigation, and improved crop varieties, and resulted in increased food output

e. The observation that average costs of production fall as output increases

f. A cycle of pesticide development, followed by pest resistance, followed by new pesticide development

g. Fertilizer produced commercially, normally with the use of fossil fuels

h. A form of soil degradation that occurs when the small amount of salts in irrigation water becomes highly concentrated on the soil surface through evaporation

i. Pesticides that target a narrower range of organisms

j. Pesticides that remain in the environment for a long time

k. Agriculture that applies the techniques of mechanization and standardization

l. The energy input per calorie of food produced

m. Fertilizers composed of organic matter from plants and animals

n. A form of soil degradation that occurs when soil remains under water for prolonged periods

o. The few individuals that are not as susceptible to a pesticide as others and may survive

p. An agricultural method that utilizes large plantings of a single species or variety

q. A pesticide that breaks down rapidly, usually in weeks or months

r. Substances, either natural or synthetic, that kill or control organisms that people consider pests

s. Pesticides that target plant species that compete with crops

Genetic engineering is revolutionizing agriculture

Key Ideas

- Genetic engineering describes the process by which scientists isolate a gene from one organism and transfer it into another.
- The benefits of genetic engineering are increased crop yield and quality, possible reduced pesticide needs, and increased profits.
- Concerns of genetic engineering include the safety of the food for human consumption, their effects on biodiversity, and how they will be regulated.

Alternatives to industrial farming methods are gaining more attention

Key Ideas

- Locations with a moderately warm climate and relatively nutrient-poor soils can lend themselves to shifting agriculture.
- Sustainable agriculture fulfills food needs while minimizing environmental harm.
- Integrated pest management tries to minimize the use of pesticides by implementing other techniques.
- Organic agriculture uses no synthetic pesticides or fertilizers.

Match These Key Terms

_____1. Conventional (industrial) agriculture

_____2. Shifting agriculture

_____3. Desertification

_____4. Nomadic grazing

_____5. Sustainable agriculture

_____6. Intercropping

_____7. Crop rotation

_____8. Agroforestry

_____9. Contour plowing

_____10. No-till agriculture

_____11. Integrated pest management

_____12. Organic agriculture

a. An agricultural technique in which crop species in a field are alternated from season to season

b. Agriculture that applies the techniques of mechanization and standardization

c. An agricultural technique in which plowing and harvesting are done parallel to the topographic contours of the land

d. An agricultural technique in which trees and vegetables are intercropped

e. An agricultural method in which two or more crop species are planted in the same field at the same time to promote a synergistic interaction

f. The transformation of arable, productive land to desert or unproductive land due to climate change or destructive land use

g. Feeding herds of animals by moving them to seasonally productive feeding grounds, often over long distances

h. An agricultural method in which farmers do not turn the soil between seasons, used as a means of reducing erosion

i. Production of crops with the goal of improving the soil each year without the use of synthetic pesticides or fertilizers

j. An agricultural method in which land is cleared and used for a few years until the soil is depleted of nutrients

k. Agriculture that fulfills the need for food and fiber while enhancing the quality of the soil, minimizing the use of nonrenewable resources, and allowing economic viability for the farmer

l. An agricultural practice that uses a variety of techniques designed to minimize pesticide inputs

Modern agribusiness includes the farming of meat and fish

Key Ideas

- High density animal farming has many negative consequences.
- Aquaculture, or fish farming, has helped to meet the growing demands for fish.

Match These Key Terms

_____1. Concentrated animal feeding operations

_____2. Fishery

_____3. Fishery collapse

_____4. Bycatch

_____5. Individual transferable quotas

_____6. Aquaculture

a. A large indoor or outdoor structure used to raise animals at very high densities

b. Farming aquatic organisms such as fish, shellfish, and seaweeds

c. A commercially harvestable population of fish within a particular ecological region

d. The unintentional catch of nontarget species while fishing

e. A fishery management program in which individual fishers are given a total allowable catch of fish in a season that they can either catch or sell

f. The decline of a fish population by 90% or more

Chapter Review

For the exam, you need to understand and be familiar with each of the ways we farm and know the benefits and consequences of each. Be sure you are comfortable with GMO's (genetically modified organisms) and fish farming.

List questions from your initial reading of the chapter

_____

_____

_____

_____

_____

_____

_____

_____

_____

_____

 **After You Read**

Short Answer

1. Distinguish between undernutrition and malnutrition and provide causes of each.

2. Where do most of the energy subsidies in modern agriculture go?

3. How can irrigation contribute to soil degradation?

4. List disadvantages and advantages of using synthetic fertilizers.

5. Explain and give examples of bioaccumulation.

6. Fill in the chart below:

Benefits of GMOs	Concerns about GMOs

7. List four sustainable agriculture practices and explain how they differ from traditional practices.

8. What are some environmental and health consequences of high-density animal farming?

Do The Math

1. If a person's food requirement per day is 2,200 kilocalories, how many kilocalories does that person need per month (assume a 30-day month)?

2. If a person only eats apples and each apple contains 53 kilocalories, how many apples will the person need to eat to get all the kilocalories they require per day? Per month?

3. If there are 6.8 billion people on the planet, how many apples would be needed per day to meet all the kilocalorie demands for the world?

Review Practice Questions: Chapters 10-11

Land Use

1. Which of the following would be an example of the tragedy of the commons?
   a. Air pollution caused by a factory
   b. Deforestation on private land
   c. A school converting their football field into a parking lot
   d. A farmer draining a lake on his land
   e. Multiple farmers letting their sheep graze on publicly owned land

2. An example of a positive externality would be
   a. lowered cost due to a sale.
   b. decreased stress after a wedding.
   c. pollution from automobile exhausts.
   d. storm prevention from a mangrove forest.
   e. overused land from cattle farming.

3. What is the concept a logger is using when he removes a particular fraction of trees on a job site in order to allow a certain amount of light to penetrate the forest floor for the younger trees?
   a. Negative externality
   b. Positive externality
   c. Tragedy of the commons
   d. Maximum sustainable yield
   e. Clear cutting

4. The majority of land in the United States is used for
   a. residential/commercial property.
   b. timber production.
   c. grassland/grazing land.
   d. forest grazing land.
   e. recreational and wildlife land.

5. The most economical way to harvest trees is to
   a. selectively cut.
   b. clear-cut.
   c. cover harvest.
   d. log.
   e. strip-cut.

6. Prescribed burns are used to
    a. destroy habitats of invasive species.
    b. lessen herbicide use.
    c. increase logging profits.
    d. open land up for housing development in a more economical way.
    e. reduce the accumulation of dead biomass.

7. One concern with tree plantations is
    a. difficulty in harvesting the trees.
    b. that only one type of rapidly growing tree species is planted.
    c. invasive species.
    d. fire control.
    e. loss of property.

8. Clear-cutting has all of the following consequences EXCEPT
    a. warmer water temperatures.
    b. erosion.
    c. expense.
    d. loss of soil nutrients.
    e. sediment buildup in nearby streams.

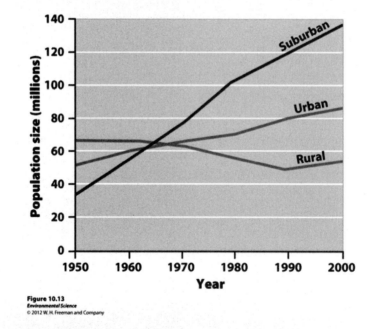

Figure 10.13
Environmental Science
© 2012 W. H. Freeman and Company

9. Which of the following is true according to the graph above?
    a. Rural populations were decreasing but have increased from 1990-2000.
    b. Suburban populations have leveled off.
    c. Approximately 85 million people live in suburban areas.
    d. Movement to urban areas is increasing at the fastest rate.
    e. All development in the United States is occurring in suburban areas.

10. According to the graph, what is the approximate percent growth of suburban populations in the United States from 1950-2000?
    a. 25%
    b. 75%
    c. 100%
    d. 285%
    e. 50%

11. Which of the following correctly summarizes urban sprawl?
    a. Large homes close together
    b. Farmland close to a river
    c. Housing and retail shops separated by miles of road
    d. Decreased traffic congestion
    e. Decreased gasoline use

12. Which of the following is NOT a characteristic of smart growth?
    a. Mixed land use
    b. Walkable neighborhoods
    c. Compact building design
    d. A variety of transportation choices
    e. Design cities without stakeholder input

13. Taking a vacant lot and building on it rather than expanding into new land outside the city is known as
    a. urban sprawl.
    b. urban blight.
    c. transit-oriented development.
    d. infill.
    e. multi-use zoning.

14. According to the World Health Organization (WHO) nearly 1/2 of the world's population is
    a. malnourished.
    b. anemic.
    c. overweight.
    d. eating feedlot beef.
    e. hunting and gathering.

15. The largest component of the human diet is
    a. grain.
    b. meat.
    c. fat.
    d. vegetables.
    e. corn.

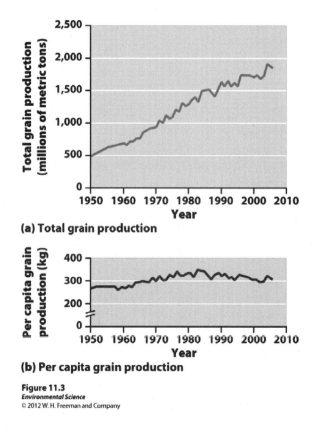

(a) Total grain production

(b) Per capita grain production

Figure 11.3
Environmental Science
© 2012 W. H. Freeman and Company

16. What can you infer from the graph above?
    a. more grain is being produced every year
    b. per capital grain production has remained completely stagnant since 1950
    c. while total grain production has risen, per capita grain production has not continued to rise
    d. per capita grain production has fallen rapidly in the last 10 year
    e. total grain production and per capita grain production have changed because of the industrial revolution.

17. Which of the following is NOT a practice of the Green Revolution?
    a. mechanization
    b. overgrazing
    c. irrigation
    d. monocropping
    e. pesticide use

18. The pesticide treadmill occurs when
    a. a farmer uses biological pest control.
    b. a homeowner mows their grass too frequently.
    c. genetically modified crops increase.
    d. pests reproduce at a faster rate.
    e. pests become resistant to pesticides and a new pesticide must be used.

19. A substance that kills insects that feed on crops but can bioaccumulate up the food chain is known as a
    a. persistent herbicide.
    b. nonpersistent herbicide.
    c. persistent insecticide.
    d. nonpersistent insecticide.
    e. genetically modified crop.

20. All of the following are benefits of genetic engineering EXCEPT
    a. greater yield.
    b. better food quality.
    c. less pesticide use.
    d. higher profits.
    e. less soil erosion.

21. Which of the following are characteristics of a genetically modified organism?
    I. a gene from one organism is transferred to a different organism
    II. they are absolutely safe for human consumption
    III. they allow a farmer to increase his revenue
    a. I only
    b. I and II
    c. II and III
    d. I and III
    e. III only

22. Desertification is happening most rapidly in what country?
    a. the United States
    b. Africa
    c. Europe
    d. Canada
    e. Japan

23. All of the following are ways to use Integrated Pest Management EXCEPT
    a. crop rotation.
    b. intercropping.
    c. using pest-resistant crop varieties.
    d. encouraging predator bugs.
    e. using persistent pesticides.

24. All of the following are characteristics of high-density animal farming EXCEPT
    a. increased strains of antibiotic-resistant microorganisms.
    b. runoff problems into area waterways.
    c. producing free-range chicken and beef.
    d. human health concerns.
    e. minimized land costs to farmers.

25. One way to alleviate some of the human-caused pressure on overexploited fisheries is to
    a. increase aquaculture practices.
    b. use long line fishing practices.
    c. encourage purse-seine fishing.
    d. use dragnets.
    e. increase the consumption of bycatch organisms.

Chapter 12: Nonrenewable Energy Resources

Summary
The major objectives of this chapter are to learn how energy is used, what nonrenewable resources are available and their uses and consequences, and how to calculate projections for future supplies of these resources.

While You Read

Nonrenewable energy accounts for most of our energy use

Key Ideas
- Fossil fuels and nuclear fuels are nonrenewable resources.
- 84% of energy consumption in the U.S. comes from fossil fuels.

Match These Key Terms

_____1. Nonrenewable energy source

_____2. Fossil fuels

_____3. Nuclear fuels

_____4. Commercial energy sources

_____5. Subsistence energy sources

a. Energy sources gathered by individuals for their own immediate needs

b. An energy source with a finite supply, primarily the fossil fuels and nuclear fuels

c. Fuels derived from radioactive materials that give off energy

d. Fuels derived from biological material that became fossilized millions of years ago

e. Energy sources that are bought and sold

Electricity is a convenient form of energy

Key Ideas
- Coal, oil, and natural gas are primary sources of energy and electricity is a secondary source of energy.
- Coal burning power plants are approximately 35% efficient.

Match These Key Terms

_____1. Energy carrier

_____2. Turbine

_____3. Electrical grid

_____4. Combined cycle

_____5. Capacity

_____6. Capacity factor

a. A device with blades that can be turned by water, wind, steam, or exhaust gas from combustion that turns a generator in an electricity-producing plant

b. A power plant that uses both exhaust gases and steam turbines to generate electricity

c. In reference to an electricity-generating plant, the maximum electrical output

d. The fraction of time a power plant operates in a year

e. A network of interconnected transmission lines that joins power plants together and links them with end users of electricity

f. Something that can move and deliver energy in a convenient, usable form to end users

Fossil fuels provide most of the world's energy

Key Ideas

- The three types of fossil fuels are coal, oil (petroleum), and natural gas.

Match These Key Terms

_____1. Coal

_____2. Petroleum

_____3. Crude oil

_____4. Oil sands

_____5. Bitumen

_____6. CTL (coal to liquid)

    a. Solid fuel formed primarily from the remains of trees, ferns, and other plant materials preserved 280 million to 360 million years ago

    b. Slow-flowing, viscous deposits of bitumen mixed with sand, water, and clay

    c. A degraded petroleum that forms when petroleum migrates to the surface of Earth and is modified by bacteria; also called tar or pitch

    d. The process of converting solid coal into liquid fuel

    e. Liquid petroleum removed from the ground

    f. A fossil fuel that occurs in underground deposits, composed of a liquid mixture of hydrocarbons, water, and sulfur

Fossil fuels are a finite resource

Key Ideas

- The Hubbert Curve is a graph that shows the point where world oil production will reach a maximum and the point where we will run out of oil.
- We have approximately 40 years of oil left and a little longer for natural gas. Coal will last us approximately 200 years and probably much longer.

Match These Key Terms

_____1. Energy intensity

_____2. Hubbert curve

_____3. Peak oil

    a. A bell-shaped curve representing oil use and projecting both when world oil production will reach a maximum and when we will run out of oil

    b. The energy use per unit of gross domestic product

    c. The point at which half the total known oil supply is used up

Nuclear energy is getting a second look

Key Ideas

- Electricity generation from nuclear energy uses the same basic process as from fossil fuels.
- An advantage of nuclear energy is that it does not produce any air pollution.
- Disadvantages of nuclear energy are the possibility of accidents and disposal of the radioactive waste.

Match These Key Terms

_____1. Fission

_____2. Fuel rods

_____3. Control rods

_____4. Radioactive waste

_____5. Becquerel (Bq)

_____6. Curie

_____7. Nuclear fusion

a. Unit that measures the rate at which a sample of radioactive material decays; 1 BQ= decay of 1 atom or nucleus per second

b. Cylindrical devices inserted between the fuel rods in a nuclear reactor to absorb excess neutrons and slow or stop the fission reaction

c. A unit of measure for radiation; 1 curie=37 billion decays per second

d. Cylindrical tubes that enclose nuclear fuel within a nuclear reactor

e. A nuclear reaction in which a neutron strikes a relatively large atomic nucleus, which then splits into two or more parts, releasing additional neutrons and energy in the form of heat

f. A reaction that occurs when lighter nuclei are forced together to produce heavier nuclei

g. Nuclear fuel that can no longer produce enough heat to be useful in a power plant but continues to emit radioactivity

Chapter Review

This chapter is extremely important to study, as you will find many questions on the AP exam based on the concepts in this chapter. Make sure you have a thorough knowledge of each energy source and the positives and negatives of each.

List questions from your initial reading of the chapter

_____

_____

_____

_____

_____

_____

_____

_____

_____

_____

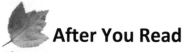 **After You Read**

Short Answer

1. What is the difference between commercial energy and subsistence energy sources?

2. What does figure 12.3 tell you about energy consumption in the United States since 1850?

3. Explain why a bus gets 1.7 MJ per passenger–kilometer while a passenger car (driver alone) gets 3.6 MJ per passenger–kilometer.

4. Summarize how a typical coal-fired electricity generation plant produces electricity.

5. Give an example of how a power company could use cogeneration to obtain greater efficiencies.

6. Summarize the process of coal formation.

7. Give 2 advantages and 2 disadvantages of using coal as an energy source.

8. Give 2 advantages and 2 disadvantages of using petroleum as an energy source.

9. Give 2 advantages and 2 disadvantages of using natural gas as an energy source.

10. Give 2 advantages and 2 disadvantages of using nuclear power as an energy source.

Do The Math

Calculating half-lives

Uranium-235 has a half live of 700 million years. How many years will it take Uranium-235 to decay to 1/8 of its original mass?

Chapter 13: Achieving Energy Sustainability

Summary
The major objectives of this chapter are to understand what renewable energy is and to be able to explain the advantages and disadvantages of each type of energy.

While You Read

What is renewable energy?
Key Ideas
- Energy resources can be renewable, nonrenewable, or potentially renewable.
- The renewable resources are wind, solar, hydroelectric, and geothermal.
- The potentially renewable resources are wood and biofuel.

Match These Key Terms

_____1. Nonrenewable energy source

_____2. Potentially renewable

_____3. Nondepletable energy source

_____4. Renewable

a. An energy source that can be regenerated indefinitely as long as it is not overharvested

b. An energy source that cannot be used up

c. An energy source with a finite supply, primarily the fossil fuels and nuclear fuels

d. In energy management, an energy source that is either potentially renewable or nondepletable

How can we use less energy?
Key Ideas
- There are many ways each individual can help to conserve energy.
- Homes and buildings can be designed for maximum energy conservation.

Match These Key Terms

_____1. Energy conservation

_____2. Tiered rate system

_____3. Peak demand

_____4. Passive solar design

_____5. Thermal inertia

a. The greatest quantity of energy used at any one time

b. The ability of a material to maintain its temperature

c. Construction designed to take advantage of solar radiation without active technology

d. The implementation of methods to use less energy

e. A billing system used by some electric companies in which customers pay higher rates as their use goes up

Biomass is energy from the Sun
Key Ideas
- The sun is the source of all fossil fuel energy as well as all renewable energy except geothermal, nuclear, and tidal energy.
- Carbon found in plants is called modern carbon and carbon found in fossil fuels is called fossil carbon.
- Wood, charcoal, and manure are examples of solid biomass that can be used for energy.
- Liquid biofuels are ethanol and biodiesel, and they can be used as substitutes for gasoline and diesel.

Match These Key Terms

_____ 1. Modern carbon	a. Carbon in fossil fuels
_____ 2. Fossil carbon	b. Alcohol made by converting starches and sugars from plant material into alcohol and carbon dioxide
_____ 3. Carbon neutral	c. A vehicle that runs on either gasoline or ethanol
_____ 4. Net removal	d. Carbon in biomass that was recently in the atmosphere
_____ 5. Ethanol	e. An activity that does not change atmospheric carbon dioxide concentrations
_____ 6. Biodiesel	f. The process of removing more than is replaced by growth, typically used when referring to carbon
_____ 7. Flex-fuel vehicles	g. A diesel substitute produced by extracting and chemically altering oil from plants

The kinetic energy of water can generate electricity

Key Ideas

- Moving water can generate electricity.
- As water moves through a dam, it turns a turbine, and energy is generated.
- There are many advantages and disadvantages of using water for energy.

Match These Key Terms

_____ 1. Hydroelectricity	a. Energy that comes from the movement of water driven by the gravitational pull of the Moon
_____ 2. Run-of-the-river	b. Hydroelectricity generation in which water is retained behind a low dam or no dam
_____ 3. Water impoundment	c. Electricity generated by the kinetic energy of moving water
_____ 4. Tidal energy	d. The storage of water in a reservoir behind a dam
_____ 5. Siltation	e. The accumulation of sediments, primarily silt, on the bottom of a reservoir

The Sun's energy can be captured directly

Key Ideas

- The Sun can be used for both passive and active solar energy.
- A photovoltaic cell can convert the Sun's energy to electricity.

Fill-in-the-Blanks

1. Active solar energy technologies capture the _____ _____ with the use of technologies.
2. Photovoltaic solar cells capture energy from the Sun as _____, not _____, and convert it directly into _____.

Earth's internal heat is a source of nondepletable energy

Key Ideas

- Heat from deep within the Earth can be used for energy.

Fill-in-the-Blanks

1. Geothermal energy is heat that comes from the natural _____ _____ of elements deep within _____.
2. Ground source heat pumps take advantage of the _____ thermal _____ of the ground.

Wind energy is the most rapidly growing source of electricity

Key Ideas
- A turbine can be used to capture wind energy and create electricity.
- Wind is nondepletable, clean, and free.

Fill-in-the-Blanks
1. Wind energy is energy generated from the _____ _____ of moving air.
2. Wind turbine converts the kinetic energy of moving air into _____.

Hydrogen fuel cells have many potential applications

Key Ideas
- Hydrogen gas can be used as a fuel source.

Fill-in-the-Blanks
1. A fuel cell is a device that operates much like a _____ _____, but with one key difference.
2. In electrolysis, an _____ _____ is applied to water to _____ it into hydrogen and _____.

How can we plan our energy future?

Key Ideas
- No single energy source is going to solve all of our energy needs.
- The U.S. will need to upgrade our existing electrical infrastructure.
- Consumers will need to demand renewable energy sources.

Fill-in-the-Blanks
1. A smart grid is an efficient, self-regulating _____ distribution network that accepts any source of _____ and distributes it automatically to _____ users.

Chapter Review

This chapter goes through the alternatives to fossil fuel and nuclear energy use. Each type of renewable energy is discussed as well as alternative gasoline options. This chapter is very important to the AP exam—make sure you have a good grasp of the concepts.

List questions from your initial reading of the chapter

_____

_____

_____

_____

**After You Read**

Short Answer

1. List 3 ways you could conserve energy.

2. What are some ways to utilize passive solar design?

3. Explain the difference between modern and fossil carbon.

4. What are some items that can be turned into ethanol and biodiesel?

5. Fill in the following chart:

Energy Resource	Advantage	Disadvantage
Liquid biofuels		
Solid biomass		
Photovoltaic solar cells		
Solar water heating systems		
Hydroelectricity		
Tidal Energy		
Geothermal energy		
Wind energy		
Hydrogen fuel cell		

Do The Math

You decide to purchase a new refrigerator. You have the choice of an Energy Star refrigerator for $2,000 or a standard unit for $1,800. The Energy Star unit costs 10 cents per hour less to run. If you buy the Energy Star unit and run it for 10 hours per day for a year, how long will it take you to recover the $200 extra cost?

Review Practice Questions: Chapters 12-13

Energy Resources and Consumption

1. If the average person in the United States uses 10,000 watts of energy, 24 hours a day for 365 days per year, how many KW of energy does the average person use in a year?
   a. 10 KW
   b. 1000 KW
   c. 3,650 KW
   d. 3,650,000 KW
   e. 87,600 KW

2. In developed countries, _____% of the world's population use _____ % of the world's energy each year.
   a. 5, 50
   b. 20, 70
   c. 5, 70
   d. 1, 50
   e. 50, 90

3. The energy source that is used most in the United States is
   a. coal.
   b. oil.
   c. natural gas.
   d. nuclear.
   e. renewables.

4. Fuel efficiency (mpg) of U.S. automobiles have
   a. decreased in the last 5 years.
   b. increased in the past 30 years.
   c. increased then decreased in the past 30 years.
   d. remained relatively stable since 1990.
   e. dropped dramatically since 1990.

5. All of the following are a part of a coal-fired electricity generation plant EXCEPT
   a. pulverizer.
   b. boiler.
   c. control rods.
   d. turbine.
   e. generator.

6. An example of cogeneration would be
    a. using steam from industrial purposes to heat buildings.
    b. using both coal and oil to create electricity.
    c. increasing nuclear power plants in major metropolitan areas.
    d. substituting anthracite coal for low grade lignite coal.
    e. operating power plants at 30% of maximum sustainable yield.

7. Put the following types of coal in order from most moisture, least heat to least moisture, most heat.
    a. peat, lignite, bituminous, anthracite
    b. peat, bituminous, lignite, anthracite
    c. anthracite, bituminous, lignite, peat
    d. bituminous, anthracite, lignite, peat
    e. bituminous, lignite, peat, anthracite

8. Which of the following statement(s) regarding petroleum is/are correct?
    I. It comes from the remains of ocean-dwelling phytoplankton that died millions of years ago
    II. It is found in porous, sedimentary rock
    III. It burns cleaner than natural gas
    a. I only
    b. II only
    c. III only
    d. I and II only
    e. I, II and III

9. Natural gas is generally found with
    a. oil.
    b. coal.
    c. both oil and coal.
    d. uranium mines.
    e. aquifers.

10. Coal supplies are expected to last for at least
    a. 5 years.
    b. 40 years.
    c. 60 years.
    d. 100 years.
    e. 200 years.

11. 1 gram of Uranium-235 contains _____ times the energy of 1 gram of coal.
    a. 20
    b. 1,000
    c. 100,000
    d. 2-3 million
    e. 1 billion

12. The difference between coal and nuclear power when it comes to how electricity is made is
    a. coal power generates steam and nuclear power does not.
    b. nuclear power uses fission to create heat to generate steam.
    c. nuclear power produces more air pollution than coal.
    d. coal is much more energy efficient than nuclear.
    e. a generator is not needed in the production of nuclear energy.

13. If a material has a radioactivity level of 100 curies and has a half-life of 10 years, how many half-lives will have occurred after 100 years?
    a. 1
    b. 4
    c. 10
    d. 1,000
    e. 25

14. Which of the following is the best example of a depletable energy source?
    a. Wind
    b. Wood
    c. Solar
    d. Geothermal
    e. Nuclear

15. If a homeowner plants a large, deciduous shade tree next to a southern window, he/she is using
    a. active solar design.
    b. photovoltaic systems.
    c. energy star technology.
    d. passive solar design.
    e. a tiered rate system.

16. In developing countries, wood is the primary resource for heating homes and cooking. What major environmental impact is associated with deforestation?
    a. Erosion
    b. Acid rain
    c. Depletion of the stratospheric ozone layer
    d. The municipal waste that is created
    e. Invasive species taking over the area

17. Most ethanol produced in the United States comes from _____ while in Brazil most ethanol comes from _____.
    a. oil deposits; corn
    b. oil sands; sugarcane
    c. corn; sugarcane
    d. biomass; wood chips
    e. wood chips; corn

18. Which of the following are environmental consequences of dams?
    I. Release of greenhouse gases
    II. Disruption to many aquatic species life cycles
    III. Accumulation of sediments in the reservoir
    a. I only
    b. II only
    c. III only
    d. I and III only
    e. II and III

19. A photovoltaic cell would be used to
    a. capture sunlight and turn it into electricity.
    b. burn biomass fuel.
    c. generate passive solar energy.
    d. generate wind power.
    e. generate electricity behind a dam.

20. The country that leads the world in the production of geothermal energy is
    a. the United States.
    b. Iceland.
    c. China.
    d. Russia.
    e. Ireland.

21. All of the following are benefits of wind power EXCEPT
    a. no pollution.
    b. silent.
    c. no greenhouse gases.
    d. unlimited source.
    e. can share land with other uses.

22. A hydrogen fuel cell takes hydrogen and oxygen and makes
    I. electricity
    II. energy
    III. water
    a. I only
    b. II only
    c. III only
    d. I and II only
    e. I, II, and III

23. A flex-fuel vehicle
    a. is known as a hybrid.
    b. uses hydrogen fuel cells for power.
    c. runs on gasoline or E-85.
    d. can drive farther on one tank of gas.
    e. runs on photovoltaic cells.

24. A disadvantage of tidal energy is that it
    a. is extremely expensive to run.
    b. generates carbon dioxide, a greenhouse gas.
    c. is aesthetically displeasing.
    d. is geographically limited.
    e. needs a storage battery to run.

25. Some energy experts say that a better system of generating electricity would be to
    a. have one large, centralized power plant.
    b. stay with our existing technology.
    c. switch to geothermal sources.
    d. have large numbers of small-scale electricity generation with fossil fuel and
       renewable energy resources.
    e. switch to hydrogen fuel cells.

Chapter 14: Water Pollution

Summary
This chapter explores the different substances that can pollute our water as well as the ways in which we clean wastewater.

 **While You Read**

Pollution can come from specific sites or broad areas

Key Ideas
- Point source pollutants come from a particular place or location while nonpoint source pollutants come from more diffuse areas.

Fill-in-the-Blanks
1. Water pollution is generally defined as the _____ of streams, rivers, lakes, oceans, or groundwater with substances produced through _____ _____ and that negatively affect _____.

2. Point sources are _____ locations such as a particular _____ that pumps its waste into a nearby stream or a sewage treatment plant that _____ its wastewater from a pipe into the ocean.

3. _____ _____ are more diffuse areas such as an entire farming region, a _____ _____ with many lawns and septic systems, or storm _____ from parking lots.

Human wastewater is a common pollutant

Key Ideas
- Human wastewater comes from sewage from toilets and grey water from bathing and washing clothes and dishes.
- Oxygen-demanding wastes can limit the amount of oxygen in the water.
- When a body of water has too many nutrients, such as nitrogen and phosphorous, the water is eutrophic.
- Human wastewater can carry a variety of illnesses such as viruses, bacteria, and parasites.

Match These Key Terms

_____1. Wastewater

_____2. Biochemical oxygen demand

_____3. Dead zones

_____4. Eutrophication

_____5. Cultural eutrophication

_____6. Indicator species

_____7. Fecal coliform bacteria

a. A species that indicates whether or not disease-causing pathogens are likely to be present

b. A phenomenon in which a body of water becomes rich in nutrients

c. Water produced by human activities including sewage from toilets and grey water from bathing and washing of clothes and dishes

d. In a body of water, areas with extremely low oxygen concentration and very little life

e. A group of microorganisms in the human intestines that can serve as an indicator species for potentially harmful microorganisms associated with contamination by sewage

f. An increase in fertility in a body of water, the result of anthropogenic inputs of nutrients

g. The amount of oxygen a quantity of water uses over a period of time at specific temperatures

We have technologies to treat wastewater from humans and livestock

Key Ideas

- Human wastewater can be treated on a small scale using septic systems or on a large scale using sewage treatment plants.
- Animal feedlots can produce a large amount of waste and can contaminate waterways.

Match These Key Terms

_____1. Septic system

_____2. Septic tank

_____3. Sludge

_____4. Septage

_____5. Leach field

_____6. Manure lagoons

a. Human-made ponds lined with rubber built to handle large quantities of manure produced by livestock

b. A component of a septic system, made up of underground pipes laid out below the surface of the ground

c. Solid waste material from wastewater

d. A large container that receives wastewater from a house as part of a septic system

e. A layer of fairly clear water found in the middle of a septic tank

f. A relatively small and simple sewage treatment system, made up of a septic tank and a leach field, often used for homes in rural areas

Heavy metals and other substances can pose serious threats to human health and the environment

Key Ideas

- Heavy metals such as lead, mercury, and arsenic can contaminate drinking water and harm human health.
- Acid deposition from burning coal can fall as rain, snow, gases, and particles, and harms plants, soil, and water.
- Pesticides, pharmaceuticals, and other manmade compounds can contaminate water.

Fill-in-the-Blanks

1. Acid deposition is when acids deposited on _____ as _____ and _____ or as gases and _____ attach to the surfaces of plants, soil, and water.

2. Polychlorinated biphenyls, or _____, represent one group of industrial _____ that has caused many _____ problems.

3. While PCBS have long been a _____, there is a growing _____ over compounds known as PBDEs (polybrominated diphenyl ethers).

Oil pollution can have catastrophic environmental impacts
Key Ideas
- Various leaks, such as oil spills from tankers and oil platforms, can harm the ocean environment.
- Oil can be cleaned up by using containment booms, vacuums to suck up the oil, chemicals to disperse the oil, and genetically modified bacteria to consume the oil.

Not all water pollutants are chemicals
Key Ideas
- Solid waste, sediment, heat (thermal pollution) and noise can also pollute water.

Fill-in-the-Blanks

1. Thermal pollution occurs when _____ _____ cause a substantial change in the _____ of _____.

2. A dramatic change in _____ temperature than can kill organisms is known as _____ _____.

A nation's water quality is a reflection of the nation's water laws and their enforcement
Key Ideas
- The Clean Water Act protects U.S. waterways.
- The Safe Drinking Water Act sets standards for U.S. drinking water.

Fill-in-the-Blanks

1. Under the _____ Drinking Water Act, the _____ is responsible for establishing _____ _____ _____ (MCL) for 77 different elements or substances in both _____ water and _____.

 **Chapter Review**

This chapter summarizes all the sources (both human and natural) that can pollute our waters. The chapter also explains the United States laws that help to protect our drinking water as well as rivers, lakes and groundwater.

List questions from your initial reading of the chapter

_____

_____

_____

_____

_____

_____

_____

_____

 **After You Read**

Short Answer

1. Define point and nonpoint source pollutants and give examples of each.

2. What are the three reasons that environmental scientists are concerned about human wastewater as a pollutant?

3. Summarize the steps in a sewage treatment plant.

4. What are some sources of lead, arsenic, and mercury?

5. Explain how DDT worked its way up the food chain and affected nonpest species.

6. How can thermal pollution harm aquatic species?

Do The Math

1. If an animal produces 60 L of manure each day and the average number of animals on a feeding operation is 460 cattle, how much manure is produced each day?

2. How much manure is produced in a week?

3. How much manure is produced in a year?

Chapter 15: Air Pollution and Stratospheric Ozone Depletion

Summary

The main objective of this chapter is to understand the major indoor and outdoor air pollutants, smog and ozone depletion. The chapter explains where pollutants come from, approaches to pollution control and prevention, and the hazards created by air pollution.

While You Read

Air pollutants are found throughout the entire global system

Key Ideas

- The major air pollutants are sulfur dioxide, nitrogen oxides, carbon oxides, particulates, lead, ozone, VOCs, and mercury.
- Primary pollutants come directly out of a car, volcano, etc., while secondary pollutants are primary pollutants that have undergone changes while in the atmosphere.

Match These Key Terms

_____1. Air pollution

_____2. Particulates

_____3. Photochemical oxidants

_____4. Smog

_____5. Photochemical smog

_____6. Sulfurous smog

_____7. Volatile organic compounds

_____8. Primary pollutants

_____9. Secondary pollutants

a. An organic compound that evaporates at typical atmospheric temperatures
b. Smog dominated by oxidants such as ozone
c. A word derived by combining the words smoke and fog
d. A primary pollutant that has undergone transformation in the presence of sunlight water, oxygen, or other compounds
e. Solid or liquid particles suspended in air
f. A class of air pollutants formed as a result of sunlight acting on compounds such as nitrogen oxides
g. Smog dominated by sulfur dioxide and sulfate compounds
h. Polluting compounds that come directly out of the smokestack, exhaust pipe, or natural emission source
i. The introduction of chemicals, particulate matter, or microorganisms into the atmosphere at high enough concentrations to harm plants, animals, and materials such as buildings, or to alter ecosystems

Air pollution comes from both natural and human sources

Key Ideas

- Natural sources of air pollution include volcanoes, lightning, forest fires, and plants.
- Anthropogenic air pollutants come from vehicles, power plants, road dust, and human-made fires.

Photochemical smog is still an environmental problem in the United States
Key Ideas
- In 2010, the EPA reported that 50 regions in the U.S. did not comply with the maximum allowable ozone concentrations in the air.
- The formation of photochemical smog is complex and still not well understood. A number of pollutants are involved and they undergo a series of complex transformations in the atmosphere.

Fill-in-the-Blanks
1. A thermal inversion occurs when a relatively _____ layer of air at mid-altitude covers a layer of _____, _____ air below.
2. The _____ inversion layer traps _____ that then accumulate beneath it.

Acid deposition is much less of a problem than it used to be
Key Ideas
- Nitrogen oxides and sulfur dioxide are released into the atmosphere as a result of numerous natural and anthropogenic combustion processes.

Pollution control includes prevention, technology, and innovation
Key Ideas
- The best way to decrease air pollution emissions is to avoid them in the first place.
- Some ways to control emissions include using catalytic converters, wet scrubbers, fluidized bed combustion, and electrostatic precipitators.

The stratospheric ozone layer provides protection from ultraviolet solar radiation
Key Ideas
- Energy from the Sun occurs at many wavelengths, including harmful high-energy ultraviolet waves, medium-energy waves (visible light), and lower-energy infrared heat waves.
- A protective layer of oxygen and ozone in the stratosphere absorbs over 99% of all incoming UV light.
- Chlorofluorocarbons (CFCs) are manmade compounds that were used in refrigeration, air conditioners, and as propellants in aerosol cans. These chemicals rose into the stratosphere and damaged the ozone layer.
- The Montreal Protocol was an agreement that 24 nations signed, pledging to reduce CFC production.

Fill-in-the-Blanks
1. The major source of _____ in the stratosphere is a class of _____ compounds known as CFCs.

Indoor air pollution is a significant hazard, particularly in developing countries

Key Ideas
- Because there is little or no ventilation in homes in developing countries, indoor air pollution from carbon monoxide and particulates pose a particular hazard.
- In developed countries, materials in the home can become sources of indoor air pollution.
- The major indoor air pollutants in developed countries are asbestos, carbon monoxide, radon, and VOCs.

Fill-in-the-Blanks
1. Asbestos is a long, thin, fibrous silicate mineral with _____ properties. The greatest health risk from asbestos has been _____ _____ such as _____ and _____ _____.
2. Sick building syndrome is a buildup of _____ compounds and pollutants in an _____ space; seen in newer buildings with good insulation and tight seals against air leaks.

Chapter Review

This chapter summarizes the causes and effects of several types of air pollutants. It describes both the primary and the secondary pollutants, the different types of smog and what causes smog, and thermal inversions and their relationship to air pollution. It goes into acid deposition and ways that we can reduce and prevent air pollution. Finally it explains in detail stratospheric ozone depletion, the benefits of the ozone layer and the ways in which it has deteriorated.

List questions from your initial reading of the chapter

_____

_____

_____

_____

_____

_____

_____

_____

_____

_____

 **After You Read**

Short Answer

1. Fill in the following chart:

Compound	Effects on Organisms
Sulfur dioxide	
Nitrogen dioxide	
Carbon monoxide	
Particulate matter	
Lead	
Ozone	
VOCs	
Mercury	
Carbon dioxide	

2. Explain the difference between photochemical smog and sulfurous smog.

3. What are some human causes of air pollution?

4. Explain the process of thermal inversion and its impact on air quality.

5. What are the effects of acid deposition?

6. What are some strategies that countries and cities have taken to lower car emissions?

7. Summarize the reactions that destroy stratospheric ozone.

8. What were sources of CFCs? Why were they banned?

9. Name some indoor air pollutants.

Do The Math
A researcher just published an article stating that CO_2 levels have increased by 8 ppm. If the level increased from 376 ppm to 384 ppm in 5 years, find the percent change of these emissions.

Chapter 16: Waste Generation and Waste Disposal

Summary
The key objective of this chapter is to understand what constitutes waste and how we deal with the problems associated with it. The 3 R's – reduce, reuse and recycle – are explained and discussed in terms of solid waste.

 **While You Read**

Humans generate waste that other organisms cannot use
Key Ideas
- The United States is the leader of what came to be known as the "throw–away society."
- The flow of solid waste that is recycled, incinerated, placed in a solid waste landfill, or disposed of in another way is called the waste stream.
- E-waste (electronic waste) is one component of waste that is small in terms of weight, but very important in terms of environmental effect and is rapidly increasing.

Match These Key Terms

_____1. Waste

_____2. Municipal solid waste(MSW)

_____3. Waste stream

a. The flow of solid waste that is recycled, incinerated, placed in a solid waste landfill, or disposed of in another way

b. Refuse collected by municipalities from households, small businesses, and institutions such as schools, prisons, municipal buildings, and hospitals

c. Material outputs from a system that are not useful or consumed

The three R's and composting divert materials from the waste stream
Key Ideas
- Three easy ways of diverting materials from the waste stream are commonly referred to as "the 3 R's": reduce, reuse, and recycle.
- Another way to reduce materials from the waste stream is to compost.

Match These Key Terms

_____1. Reduce, Reuse, Recycle

_____2. Source reduction
_____3. Reuse

_____4. Recycling

_____5. Closed-loop recycling
_____6. Open-loop recycling
_____7. Compost

a. Organic matter that has decomposed under controlled conditions to produce an organic-rich material that enhances soil structure, cation exchange capacity, and fertility

b. Recycling a product into the same product

c. A popular phrase promoting the idea of diverting materials from the waste stream

d. The reduction of waste through minimizing the use of materials destined to become municipal solid waste from the early stages of design and manufacture

e. Using a product or material that was intended to be discarded

f. Recycling one product into a different product

g. The process by which materials destined to become municipal solid waste are collected and converted into raw material that is then used to produce new objects

Currently, most solid waste is buried in landfills or incinerated

Key Ideas

- The U.S. used to put waste in open dumps.
- Sanitary landfills are designed to hold the waste with as little contamination of the surrounding environment as possible.
- Incineration is the act of burning waste to reduce volume and mass. It is sometimes used to generate electricity or heat.

Match These Key Terms

_____1. Leachate

_____2. Sanitary landfills

_____3. Tipping fee

_____4. Siting

_____5. Incineration

_____6. Ash

_____7. Bottom ash

_____8. Fly ash

a. The residual nonorganic material that does not combust during incineration

b. Residue collected at the bottom of the combustion chamber in a furnace

c. The residue collected from the chimney or exhaust pipe of a furnace

d. Liquid that contains elevated levels of pollutants as a result of having passed through municipal solid waste or contaminated soil

e. The process of burning waste materials to reduce volume and mass, sometimes to generate electricity or heat

f. Engineered ground facilities designed to hold municipal solid waste with as little contamination of the surrounding environment as possible

g. The designation of a landfill location, typically through a regulatory process involving studies, written reports, and public hearings

h. A fee charged for disposing of material in a landfill or incinerator

Hazardous waste requires special means of disposal

Key Ideas

- The majority of hazardous waste comes from industry.
- In the U.S., hazardous waste must be treated before disposal.
- The Superfund (CERCLA) law authorized the federal government to respond to any substance that may pose a threat to human health or the environment.
- Brownfields are sites that are not as hazardous as Superfund sites but are still contaminated.

Match These Key Terms

_____1. Hazardous waste

_____2. Superfund

_____3. Brownfields

a. CERCLA, a 1980 U.S. federal act that imposes a tax on the chemical and petroleum industries, funds the cleanup of abandoned and nonoperating hazardous waste sites, and authorizes the federal government to respond directly to the release or threatened release of substances that may pose a threat to human health or the environment

b. Waste material that is dangerous or potentially harmful to humans or ecosystems

c. Contaminated industrial or commercial sites that may require environmental cleanup before they can be redeveloped or expanded

There are newer ways of thinking about solid waste

Key Ideas

- Every method of waste disposal will have adverse environmental effects.
- It is important to analyze the life-cycle of any material.
- Integrated Waste Management uses the ideas of reduce, reuse, recycle.

Fill-in-the-Blanks

1. Life-cycle analysis is an important systems _____ that looks at the materials

   _____ and _____ throughout the lifetime of a product.

2. Integrated Waste Management, which employs several _____,

   management, and _____ _____ in order to reduce the

   environmental _____ of MSW.

 **Chapter Review**

This chapter is about building an awareness of our throwaway society and how we deal with all of the trash we create. Waste goes into landfills, is recycled or is incinerated. This chapter gives the positives and negatives of each solution to the problem of trash disposal and reviews the laws that have been written to protect us from both household waste and hazardous waste.

List questions from your initial reading of the chapter

_____

_____

_____

_____

_____

_____

_____

_____

_____

_____

 **After You Read**

1. Summarize the composition of municipal solid waste in the U.S.

2. List 5 parts of a sanitary landfill.

3. What are some problems with landfills and how can these problems be prevented?

4. What are some problems with incineration of waste?

5. Summarize the RCRA and CERCLA laws.

6. Explain Integrated Waste Management.

Do The Math

1. The annual precipitation at a landfill is 250mm per year, and 50% of this water runs off the landfill. If the landfill has a surface area of 10,000 m^2 and the leachate collection system is 80% effective, calculate the volume of water in cubic meters that infiltrates the landfill per year.

2. How much volume of leachate in m^3 is treated per year?

Chapter 17: Human Health and Environmental Risks

Summary

The main objectives of this chapter are to understand human health risks, infectious diseases, toxic chemicals and risk analysis.

While You Read

Human health is affected by a large number of risk factors

Key Ideas

- The three major categories of risks that can harm human health are physical, biological, and chemical.
- Diseases can be spread from person to person.
- Malnutrition and poor sanitation are risk factors for people living in developing nations.

Match These Key Terms

_____1. Disease
_____2. Infectious disease
_____3. Chronic diseases
_____4. Acute diseases

a. Diseases that rapidly impair the functioning of an organism
b. Diseases that slowly impairs functioning of an organism
c. Any impaired function of the body with a characteristic set of symptoms
d. A disease caused by a pathogen

Infectious disease have killed large numbers of people

Key Ideas

- The plague, malaria, and tuberculosis are three historically infectious diseases.
- HIV/AIDS, Ebola, mad cow disease, bird flu, and West Nile virus are some emergent infectious diseases that have become prevalent in the last 20 years.

Match These Key Terms

_____1. Epidemic
_____2. Pandemic
_____3. Plague
_____4. Malaria
_____5. Tuberculosis
_____6. Emergent infectious diseases
_____7. AIDS
_____8. HIV
_____9. Ebola hemorrhagic fever
_____10. Mad cow disease
_____11. Prions
_____12. Bird flu
_____13. West Nile virus

a. A virus that lives in hundreds of species of birds and is transmitted by mosquitoes
b. A highly contagious disease caused by the bacterium *Mycobacterium tuberculosis* that primarily infects the lungs
c. Infectious disease caused by the bacterium *Yersinia pestis*, carried by fleas
d. An epidemic that occurs over a large geographic region
e. Small, beneficial proteins that occasionally mutate into a pathogen
f. A disease in which prions mutate into deadly pathogens and slowly damage a cow's nervous system
g. An infectious disease caused by one of several species of protists in the genus *Plasmodium*
h. An infectious disease with high death rates caused by the Ebola virus
i. An infectious disease that has not been previously described or has not been common for at least 20 years
j. A situation in which a pathogen causes a rapid increase in disease
k. A virus that causes AIDS
l. An avian influenza caused by the H1N1 virus
m. An infectious disease caused by HIV

Toxicology is the study of chemical risks
Key Ideas
- The 5 types of chemical toxins are neurotoxins, carcinogens, teratogens, allergens, and endocrine disruptors.

Match These Key Terms

_____1. Neurotoxins

_____2. Carcinogens

_____3. Mutagens

_____4. Teratogens

_____5. Allergens

_____6. Endocrine disruptors

a. Chemicals that interfere with the normal functioning of hormones in an animal's body

b. Chemicals that cause cancer

c. Carcinogens that cause damage to the genetic material of a cell

d. Chemicals that cause allergic reactions

e. Chemicals that disrupt the nervous system of animals

f. Chemicals that interfere with the normal development of embryos or fetuses

Scientists can determine the concentrations of chemicals that harm organisms
Key Ideas
- Scientists must determine the risk a chemical poses to organisms.
- The way in which a person might come into contact with a chemical is important.
- Some chemicals have worked their way up the food chain.

Match These Key Terms

_____1. Dose-response studies

_____2. Acute studies

_____3. LD50

_____4. Sublethal effects

_____5. ED50

_____6. Chronic studies

_____7. Epidemiology

_____8. Retrospective studies

_____9. Prospective studies

_____10. Synergistic interactions

_____11. Routes of exposure

_____12. Solubility

_____13. Biomagnification

_____14. Persistence

a. Risks that cause more harm together than expected based on separate individual risks

b. How well a chemical dissolves in a liquid

c. Studies that monitor people who have been exposed to an environmental hazard at some time in the past

d. Experiments that expose organisms to an environmental hazard for a short duration

e. The increase in chemical concentration in animal tissues as the chemical moves up the food chain

f. Studies that expose organisms to different amounts of a chemical and then observes a variety of possible responses, including mortality or changes in behavior or reproduction

g. The effective dose of a chemical that causes 50% of the individuals in a dose-response study to display a harmful, but nonlethal, effect

h. The lethal dose of a chemical that kills 50% of the individuals in a dose-response study

i. The study of causes of illness and disease in the populations of humans and other organisms

j. The length of time a chemical remains in the environment

k. Experiments that expose organisms to an environmental hazard for a long duration

l. The effects of an environmental hazard that are not lethal, but which may impair an organism's behavior, physiology, or reproduction

m. Studies that monitor people who might become exposed to harmful chemicals in the future

n. The way in which an individual might come into contact with an environmental hazard

Risk analysis helps us assess, accept, and manage risk

Key Ideas

- Qualitative risk assessment is how we evaluate a given risk by using our judgment.
- Quantitative risk assessment brings together tremendous amounts of data to determine risk.
- The Stockholm Convention produced a list of 12 chemicals known as the "dirty dozen," which should be banned or phased out.

Match These Key Terms

_____1. Environmental hazard

_____2. Innocent-until-proven-guilty principle

_____3. Precautionary principle

_____4. Stockholm Convention

_____5. REACH

a. A 2007 agreement among the nations of the European Union about regulation of chemicals

b. A principle based on the philosophy that action should be taken against a plausible environmental hazard

c. A 2001 agreement among 127 nations concerning 12 chemicals to be banned, phased out, or reduced

d. A principle based on the philosophy that a potential hazard should not be considered an actual hazard until the scientific data definitively demonstrate that it actually causes harm

e. Anything in the environment that can potentially cause harm

 **Chapter Review**

This chapter reviews numerous chemical and biological risks to the environment. Some of the risks are caused by humans while others occur naturally. This information is probably new to you and not a review of concepts you have learned before in other science courses, so be sure to spend extra time with it before the AP exam if you need to.

List questions from your initial reading of the chapter

_____

_____

_____

_____

_____

_____

_____

_____

After You Read

1. What are some health risks for people living in a developing nation? A developed nation?

2. Fill out the following chart:

Disease	How Disease is Spread
Plague	
Malaria	
Tuberculosis	
HIV/AIDS	
Ebola hemorrhagic fever	
Mad cow disease	
Bird flu	
West Nile virus	

3. What are the different pathways of transmitting pathogens?

4. Fill out the following chart:

Chemical	Source	Type	Effect
Lead			
Mercury			
Arsenic			
Asbestos			
PCBs			
Radon			
Vinyl chloride			
Alcohol			
Atrazine			
DDT			
Phthalates			

5. Explain the difference between LD50 and ED50.

6. How did DDT move up the food chain and harm predatory birds?

7. What factors are included in risk analysis?

Review Practice Questions: Chapters 14-17

Pollution

1. Which of the following would be an example of a point source water pollutant?
    a. Agricultural lands
    b. Animal feedlots
    c. Runoff from parking lots
    d. Factory effluent
    e. Residential lawns

2. When sewage contaminates a body of water, it can lead to a lower dissolved oxygen level in the water. This is because
    a. the sewage has put a high BOD on the water.
    b. the water has too many fish and other organisms.
    c. the water was unsafe to begin with and the sewage magnified this.
    d. a septic tank is leaking in the area.
    e. the lake was oligotrophic.

3. The World Health Organization estimates that _____ of the world's population does not have access to sufficient supplies of safe drinking water.
    a. 1 out of every 100
    b. 1 out of every 50
    c. 1 out of every 25
    d. 1 out of every 10
    e. 1 out of every 6

Match the following steps in the sewage treatment process
    4. _____ biological   a. solid waste materials settle out
    5. _____ chemical   b. bacteria break down 85-90% of organic matter
    6. _____ mechanical   c. chlorine, ozone, or ultraviolet light are used

7. All of the following are examples of primary pollutants EXCEPT
    a. carbon monoxide.
    b. ozone.
    c. carbon dioxide.
    d. sulfur dioxide.
    e. volatile organic compounds.

8. Which of the following is a corrosive gas that comes primarily from burning coal?
   a. Carbon monoxide
   b. Ozone
   c. Carbon dioxide
   d. Sulfur dioxide
   e. Volatile organic compounds

9. Which of the following chemical(s) cause acid deposition?
   I. sulfur dioxide
   II. nitrogen oxides
   III. carbon dioxide
   a. I only
   b. II only
   c. III only
   d. I and II
   e. I, II, and III

10. A coal burning power plant that uses an electrostatic precipitator is designed to remove
    a. sulfur dioxide.
    b. nitrogen dioxide.
    c. particulate matter.
    d. carbon dioxide.
    e. methane.

11. A thermal inversion causes severe pollution events. Thermal inversions
    a. occur when a warm inversion layer traps emissions that then accumulate beneath it.
    b. occur when a cold inversion layer traps emissions that then accumulate beneath it.
    c. occur when a warm inversion layer traps emissions that then accumulate above it.
    d. occur when cold inversion layer traps emissions that then accumulate above it.
    e. are generally worse in mid-latitudes.

12. Which of the following statements about stratospheric ozone depletion is correct?
    a. Ozone depletion is the result of automobiles and coal fired power plants
    b. Ozone depletion occurs when infrared heat is trapped near the earth
    c. Ozone depletion is causing melting of our polar ice caps
    d. Ozone depletion is causing respiratory disease to increase
    e. Ozone depletion allows more ultraviolet waves to pass through to the troposphere

13. Which of the following chemicals is responsible for destroying the stratospheric ozone layer?
    a. Carbon
    b. Chlorine
    c. Methane
    d. Sulfur dioxide
    e. Fluorine

14. Which of the following items makes up most of the municipal solid waste stream in the US?
    a. Paper
    b. Plastic
    c. Yard trimmings
    d. Food scraps
    e. Wood

15. Which of the following uses the least amount of energy?
    a. Reduce
    b. Reuse
    c. Closed-loop recycling
    d. Open-loop recycling
    e. Incineration

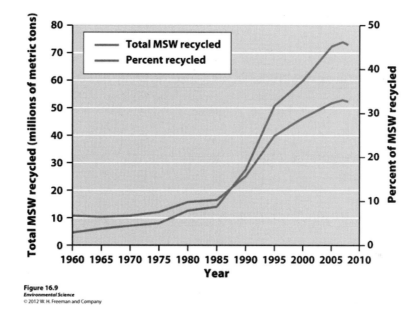

Figure 16.9
Environmental Science
© 2012 W. H. Freeman and Company

16. Using the graph above, what is the approximate percent change of materials that were recycled in 2008 compared to 1960?
    a. 10%
    b. 25%
    c. 75%
    d. 100%
    e. 380%

17. All of the following are consequences of incineration EXCEPT
    a. fly ash is produced.
    b. fossil fuels are burned transporting the waste.
    c. leachate is produced.
    d. air pollution can occur.
    e. metals and other toxins may be released.

18. Which of the following laws was created because of Love Canal?
    a. RCRA
    b. Brownfields
    c. ESA
    d. CERCLA
    e. FIFRA

19. Integrated Waste Management employs
    I. source reduction
    II. recycling
    III. composting
    a. I only
    b. II only
    c. III only
    d. I and II only
    e. I, II, and III

20. The number one leading health risk in high-income countries is
    a. automobiles.
    b. tobacco use.
    c. obesity.
    d. low fruit and vegetable intake.
    e. high cholesterol.

21. All of the following are infectious diseases EXCEPT
    a. cancer.
    b. plague.
    c. tuberculosis.
    d. HIV/AIDS.
    e. Ebola.

22. Which of the following chemicals is a neurotoxin?
    a. Arsenic
    b. Asbestos
    c. Lead
    d. DDT
    e. Radon

23. Which of the following experiment results correctly describes the effects of a LD50 dose of a chemical?
    a. 1 out of every 50 rats die
    b. 50 out of every 100 rats die
    c. 1 out of every 50 rats gets sick
    d. 50 out of every 100 rats get sick
    e. All rats die

24. Which of the following would be an example of a synergistic interaction?
    a. Increased riders in a car leads to increased death
    b. Smoking causes lung disease
    c. Tuberculosis is spread among the poor
    d. Recreational drug use can lead to HIV/AIDS
    e. Asbestos exposure along with smoking can exasperate lung cancer

25. DDT exposure to plankton then is transferred to small fish, to larger fish, and eventually to predatory birds. This is an example of
    a. a food chain.
    b. a synergistic effect.
    c. risk assessment.
    d. biomagnification.
    e. teratogens.

Chapter 18: Conservation of Biodiversity

Summary
This chapter addresses biodiversity with a focus on conservation. It also looks at causes of declining biodiversity and the laws created in order to protect species.

 **While You Read**

We are in the midst of a sixth mass extinction
Key Ideas

- Instrumental values are things such as food, medicine, and building materials that humans need or use. Intrinsic values provide no direct benefit to people, but follow the belief that individuals, species, and ecosystems are valuable in themselves and should be preserved.
- We are in the midst of a 6th extinction caused by humans.
- There has been a decline of genetic diversity in wild organisms and crops.

Match These Key Terms

_____1. Extinction   a. The death of the last member of a species

_____2. Inbreeding depression   b. At serious risk of extinction

_____3. Endangered   c. A genetic phenomenon in which individuals with similar genotypes breed with each other and produce offspring with an impaired ability to survive and reproduce

Declining biodiversity has many causes
Key Ideas

- For most species, the greatest cause of decline and extinction is habitat loss.
- Native species live in their historical ranges and exotic species live outside their historical range.
- Species can be overharvested and become endangered.
- Pollution and climate change can lead to loss of biodiversity.

Match These Key Terms

_____1. Native species   a. A species that spreads rapidly across large areas

_____2. Alien species   b. A species that lives in a historical range, typically where it has lived for thousands or millions of years

_____3. Invasive species   c. A 1973 treaty formed to control the international trade of threatened plants and animals

_____4. Lacey Act   d. A species living outside its historical range

_____5. CITES   e. A list of worldwide threatened species maintained by the International Union for Conservation of Nature

_____6. Red List   f. A U.S. act that prohibits interstate shipping of all illegally harvested plants and animals

The conservation of biodiversity often focuses on single species
Key Ideas
- There are two approaches to protecting biodiversity: the single-species approach and the ecosystem approach

Fill-in-the-Blanks
1. The Convention on Biological Diversity is a 1992 _____ treaty formed to help protect _____.

The conservation of biodiversity sometimes focuses on protecting entire ecosystems
Key Ideas
- Some factors that must be taken into account when protecting species are the size and shape of the protected area, the connectedness to other protected areas, and how best to incorporate conservation while recognizing the need for sustainable habitat for humans.

Fill-in-the-Blanks
1. Edge habitat occurs where _____ different communities come together, typically forming an _____ transition, such as where a grassy _____ meets a _____.
2. Biosphere reserves are _____ areas consisting of _____ that vary in the amount of permissible _____ impact.

 Chapter Review

This chapter summarizes many of the ideas from the chapters that have preceded it. It looks at how human actions have affected the animals and plants with which we share the planet. Make sure you are familiar with all the laws that this chapter covers because they often show up on the AP test.

List questions from your initial reading of the chapter

_____

_____

_____

_____

_____

_____

_____

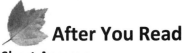 **After You Read**

Short Answer

1. What are the five categories concerning the status of a species?

2. How do invasive species threaten biodiversity? Give an example of an invasive species.

3. Summarize the aims of following laws and treaties:
   Lacey Act

   CITES

   Marine Mammal Protection Act

   Endangered Species Act

   Convention on Biological Diversity

4. Summarize the idea behind biosphere reserves.

5. What are some of the reasons behind biodiversity decline?

6. How can we reduce species decline?

Chapter 19: Global Change

Summary
The principle objectives of this chapter are to understand the concepts of global change, global climate change, and global warming. You should have a firm grasp on the ways human actions affect global change and the impact of these actions on the environment.

While You Read

Global change includes global climate change and global warming
Key Ideas
- Change that occurs in the chemical, biological, and physical properties of the planet is called global change.
- One type of global change is climate change.

Match These Key Terms

_____1. Global change	a. Change that occurs in the chemical, biological, and physical properties of the planet
_____2. Global climate change	b. The warming of the oceans, landmasses, and atmosphere of Earth
_____3. Global warming	c. Changes in the climate of Earth

Solar radiation and greenhouse gases make our planet warm
Key Ideas
- The ultimate source of almost all energy on Earth is the Sun.
- The gases in the atmosphere that absorb infrared radiation are known as greenhouse gases.
- Much like how a gardener's greenhouse traps heat from the sun, the greenhouse effect keeps Earth warm enough to be habitable.

Fill-in-the-Blanks
1. The greenhouse warming potential of a _____ estimates how much a _____ of any _____ can contribute to global warming over a period of _____ years relative to a molecule of _____.

Sources of greenhouse gases are both natural and anthropogenic
Key Ideas
- Natural greenhouse gases come from volcanic eruptions, decomposition, digestion, denitrification, evaporation, and evapotranspiration.
- The most significant anthropogenic sources of greenhouse gases are the burning of fossil fuels, agriculture, deforestation, landfills, and industrial production of chemicals.

Changes in CO2 and global temperatures have been linked for millennia

Key Ideas

- One way to determine if human activities are causing global warming is to look at gas concentrations and temperatures from the past and compare them to gas concentrations and temperatures in the present day.
- The IPCC concluded that CO_2 is an important greenhouse gas that can contribute to global warming.
- As more nations industrialize and more fossil fuels are burned, carbon dioxide levels are increasing.
- Data from around the globe shows that temperatures have increased by 0.8 degrees Celsius from 1880-2009.
- You can indirectly measure climate change by looking at organisms that have been preserved for millions of years and chemically analyzing ice from long ago.
- Computer models are helping us predict future CO_2 changes.

Feedbacks can increase or decrease the impact of climate change

Key Ideas

- Positive feedbacks amplify change while negative feedbacks restrict change.

Global warming has serious consequences for the environment and organisms

Key Ideas

- Warmer temperatures are causing the melting of polar ice caps, glaciers, and permafrost, resulting in rising sea levels.
- Predicted effects for the future are increased frequency of heat waves, reduced cold spells, altered precipitation patterns and storm intensity, and shifting ocean currents.
- The warming planet is also having an effect on living organisms.

The Kyoto Protocol addresses climate change at the international level

Key Ideas

- In 1997, representatives of the nations of the world convened in Kyoto, Japan to discuss how best to control the emissions contributing to global warming.

Fill-in-the-Blanks

1. An international agreement to reduce global emissions of greenhouse gases is known as the

    _____   _____.

2. Carbon sequestration is the approach of taking _____ out of the

    atmosphere.

 **Chapter Review**

This chapter looks at the causes—both natural and human—and consequences of climate change. It also explains the greenhouse effect and the history of climate change. The chapter ends with the summary from the IPCC in 2007 (Table 19.2, page 540) that reviews methods countries are looking at to reverse the trend. It is very important to know what causes climate change, the chemicals involved in the process, and the environmental effects of a warmer climate.

List questions from your initial reading of the chapter

_____

_____

_____

_____

_____

_____

_____

_____

**After You Read**

Short Answer

1. Summarize the greenhouse effect.

2. List the five greenhouse gases and their relative impact on the greenhouse effect. Include their global warming potential and how long they remain in the atmosphere.

3. Why do carbon dioxide levels change seasonally?

4. How have scientists used ice cores to determine past climate patterns? How can these patterns help scientists make predictions about future climate change?

5. What are the two explanations for warming temperatures on Earth?

6. Explain carbon sequestration.

Do The Math

1. In 2010, the concentration of carbon dioxide in the atmosphere was 390ppm. If the annual rate of carbon dioxide increase is 1.4 ppm, what concentration of carbon dioxide do you predict for the year 2050?

2. If the annual rate of carbon dioxide increases from 1.4ppm to 1.9ppm, what will the concentration of carbon dioxide be in the year 2050?

Chapter 20: Sustainability, Economics, and Equity

Summary

The principle objectives of this chapter are to understand the effect economic activity has on the environment and on human well-being. The laws designed to protect the environment and human beings are discussed. This chapter also discusses the connections between economic costs and environmental protection together with social ramifications.

 **While You Read**

Sustainability is the ultimate goal of sound environmental science and policy

Key Ideas

- Something is sustainable when it meets the needs of the present generation without compromising the ability of future generations to meet their own needs

Fill-in-the-Blanks

1. Well-being is the _____ of being _____, happy and _____.

Economics studies how scarce resources are allocated

Key Ideas

- The cost or impact of a good or service on people and the environment is known as an externality.
- The Kuznets curve suggests that as per capita income in a country increases, environmental degradation first increases and then decreases.

Match These Key Terms

_____1. Economics

_____2. GPI

_____3. Technology transfer

_____4. Leapfrogging

_____5. Microlending

a. A measurement of the economy that considers personal consumption, income distribution, levels of higher education, resource depletion, pollution, and the health of the population

b. The situation in which less developed countries use newer technology without first using the precursor technology

c. The practice of loaning small amounts of money to help people in less developed countries start small businesses

d. The phenomenon of less developed countries adopting technological innovations that originated in wealthy countries

e. The study of how humans either as individuals or as companies allocate scarce resources in the production, distribution, and consumption of goods and services

Economic health depends on the availability of natural capital and basic human welfare

Key Ideas

- Capital is divided into three categories: natural, human, and manufactured.

Match These Key Terms

_____1. Natural capital

_____2. Human capital

_____3. Manufactured capital
_____4. Market failure
_____5. Environmental economics
_____6. Ecological economics

_____7. Valuation

a. The economic situation that results when the economic system does not appropriately account for all costs

b. A subfield of economics that examines costs and benefits of various polices and regulations related to environmental degradation

c. Natural resources of Earth, such as air, water, and minerals

d. Human knowledge, potential, and abilities

e. The study of economics as a component of ecological systems

f. The practice of assigning monetary value to seemingly intangible benefits and natural capital

g. All goods and services that humans produce

Agencies, laws, and regulations are designed to protect our natural and human capital

Key Ideas

- Three types of environmental worldviews are human-centered, life-centered, and Earth-centered.
- Major world agencies that deal with environmental problems are the United Nations, the World Bank, the World Health Organization, and the United Nations Development Program.

Match These Key Terms

_____1. Environmental worldviews
_____2. Stewardship

_____3. Biocentric

_____4. Ecocentric

_____5. United Nations

_____6. United Nations Environment Programme

_____7. World Health Organization
_____8. United Nations Development Programme

_____9. Environmental Protection Agency

_____10. OSHA

_____11. Department of Energy

_____12. World Bank

_____13. Anthropocentric

a. A worldview that focuses on human welfare and well-being

b. A worldview that considers human beings to be just one of many species on Earth, all of which have equal intrinsic value

c. Views that encompass how people think the world works, how they view their role in it, and what they believe to be proper behavior regarding the environment

d. A worldview that places equal value on all living organisms and the ecosystems in which they live

e. A U.S. government agency that creates federal policy and oversees enforcement of regulations related to the environment, including science, research, assessment, and education

f. A U.S. government agency created in 1977 with the goal of advancing the energy and economic security of the U.S.

g. Careful and responsible management of Earth and its resources

h. A program of the U.N. responsible for gathering environmental information and conducting research and assessing environmental problems

i. An institution dedicated to promoting dialogue among countries with the goal of maintaining world peace

j. A program of the U.N. that works to improve living conditions through economic development

k. A group within the U.N. responsible for human health, including combating the spread of infectious diseases and health issues related to natural disasters

l. An international organization that provides technical and financial assistance to help reduce poverty and promote growth, especially in the world's poorest countries

m. A U.S. federal agency responsible for the enforcement of health and safety regulations in the workplace

There are several approaches to measuring and achieving sustainability

Key Ideas

- The human development index combines three basic measures of human status.
- The five basic steps in a policy cycle are problem identification, policy formulation, policy adoption, policy implementation, and policy evaluation.

Match These Key Terms

_____1. HDI

_____2. HPI

_____3. Command-and-control approach

_____4. Incentive-based approach

_____5. Green tax

_____6. Triple bottom line

a. A program that constructs financial and other incentives for lowering emissions, based on profits and benefits

b. An approach to sustainability that advocates consideration of economic, environmental, and social factors in decisions about business, the economy, the environment, and development

c. A tax placed on environmentally harmful activities or emissions

d. Developed by the U.N. to investigate the proportion of a population suffering from deprivation in a country with a high HDI

e. A strategy for pollution control that involves regulations and enforcement mechanisms

f. A measure of economic well-being that combines life expectancy, knowledge, education, and standard of living as shown in GDP per capita and purchasing power

Two major challenges of our time are reducing poverty and stewarding the environment

Key Ideas

- Approximately 1 billion people live in unsanitary conditions.
- Dr. Wangari Maathai founded the Green Belt Movement.
- Environmental equity is the fair distribution of Earth resources.
- Environmental justice is a social movement that examines whether there is equal enforcement of environmental laws and elimination of disparities.

 **Chapter Review**

This chapter is about balancing environmental, economic, and social goals and using the tools of economics to help make environmental decisions. Make sure you are familiar with supply and demand and gross domestic product (GDP) as these topics frequently appear on the AP exam.

List questions from your initial reading of the chapter

_____

_____

_____

_____

**After You Read**

Short Answer

1. Explain supply and demand.

2. Summarize the concepts of microloans and the potential benefits to their recipients. Why is this discussed in environmental science courses?

3. What are the eight Millennium Development Goals?

4. What was Dr. Wangari Maathai's major contribution to the field of environmental science?

Review Practice Questions: Chapters 18-20

Global Change and a Sustainable Future

1. Which greenhouse gas traps the majority of outgoing infrared radiation?
   a. Methane
   b. Water vapor
   c. Carbon dioxide
   d. Sulfur dioxide
   e. Nitrogen dioxide

2. All of the following are causes of biodiversity decline EXCEPT
   a. habitat loss.
   b. alien species.
   c. overharvesting.
   d. pollution.
   e. banning lead in gasoline.

3. The greatest cause of biodiversity decline is
   a. habitat loss.
   b. alien species.
   c. overharvesting.
   d. pollution.
   e. banning lead in gasoline.

4. All of the following describe invasive species EXCEPT
   a. they spread rapidly.
   b. they have no natural enemies.
   c. they cause harmful effects on native species.
   d. they are never introduced by humans.
   e. they can outcompete native species.

5. All of the following are acts of legislation or treaties to help control biodiversity decline EXCEPT
   a. Endangered Species Act
   b. Marine Mammal Protection Act
   c. CITES agreement
   d. Whaling Commission Protective Act
   e. Convention on Biological Diversity

6. The greenhouse effect happens when
   a. the ozone layer is decreased.
   b. infrared radiation is absorbed and emitted back to Earth.
   c. UV light is trapped at earth's surface.
   d. greenhouse gasses emit UV light.
   e. CFC's destroy the stratospheric ozone layer.

7. Which of the following chemicals has the greatest greenhouse warming potential?
   a. Water vapor
   b. Carbon dioxide
   c. Chlorofluorocarbons
   d. Methane
   e. Nitrous oxide

8. Which greenhouse gas comes from automobiles?
   a. Nitrous oxide
   b. Methane
   c. Sulfur dioxide
   d. Ozone
   e. Chlorofluorocarbons

9. Which of the following are anthropogenic causes of greenhouse gasses?
   I. Volcanic eruptions
   II. Decomposition
   III. Coal burning power plants
   a. I only
   b. II only
   c. III only
   d. I and II
   e. I, II and III

10. Which of the following is not an anthropogenic source of greenhouse gases?
    a. Burning of fossil fuels
    b. Agriculture
    c. Deforestation
    d. Landfills
    e. Asbestos

11. Which source produces greenhouse gases and can also increase levels of mercury in the environment?
    a. Refrigerators
    b. Coal
    c. Automobiles
    d. Landfills
    e. Agricultural practices

12. All of the following sources can produce methane EXCEPT
   a. livestock.
   b. sewage treatment plants.
   c. cement manufacturing.
   d. wetlands.
   e. termites.

13. Which country produces the most carbon emissions?
   a. The United States
   b. China
   c. Australia
   d. India
   e. England

14. Why do levels of carbon dioxide in our atmosphere have seasonal variations?
   a. More fossil fuels are burned in the winter for heat.
   b. More fossil fuels are burned in the summer for travel.
   c. Livestock production increases each spring.
   d. Photosynthesis varies each season.
   e. Landfills increase in size as more trash is produced during summer month.

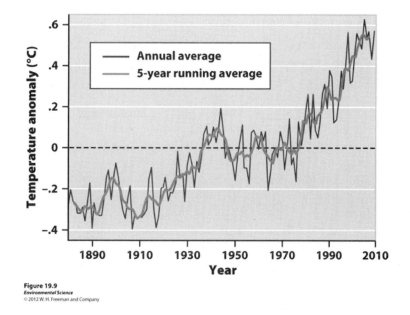

Figure 19.9
Environmental Science
© 2012 W. H. Freeman and Company

15. According to the graph above,
   a. temperature has been increasing steadily since 1975.
   b. from 1890 to 2010, average temperatures over 5 year periods do not fluctuate.
   c. the highest temperature recorded was in 1940.
   d. temperature has been declining recently.
   e. the coldest temperature recorded was in 1920.

16. All of the following are impacts that warmer temperatures can have on the environment EXCEPT
    a. melting of polar ice caps.
    b. rising sea levels.
    c. heat waves.
    d. increased storms.
    e. larger hole in the stratospheric ozone layer.

17. What is the name of the agreement that worked to control the emissions that contribute to global warming?
    a. Montreal Protocol
    b. Kyoto Protocol
    c. Convention on Biological Diversity
    d. Clean Air Act
    e. Climate Change Act

18. Storing carbon in agricultural soils to return atmospheric carbon to longer-term storage in the form of plant biomass is known as
    a. agricultural restoration.
    b. soil reclamation.
    c. carbon sequestration.
    d. carbon dioxide scrubber.
    e. greenhouse gas depletion.

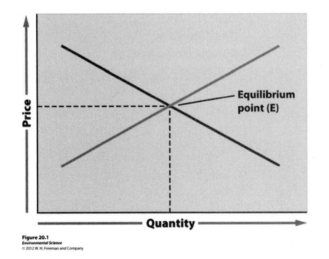

Figure 20.1
Environmental Science
© 2012 W. H. Freeman and Company

19. According to the graph above, what is point where supply meets demand?
    a. Where price exceeds quantity
    b. Where quantity exceeds price
    c. When supply rises above demand
    d. Where price and quantity meet
    e. To the right of the equilibrium point

20. When purchasing a desk, you are paying for the materials, labor, shipping, etc., but you are not paying for the environmental cost of air pollution that was generated in making the desk. This environmental cost is known as a(n)
    a. externality.
    b. environmental adjustor.
    c. market price.
    d. supply side curve.
    e. genuine progress indicator.

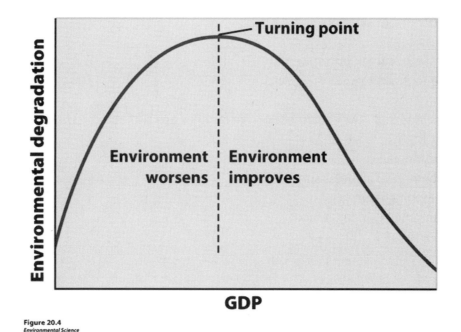

Figure 20.4
Environmental Science
© 2012 W. H. Freeman and Company

21. What does the Kuznet curve (above) explain?
    a. As per capita income decreases, environmental degradation decreases.
    b. As per capita income decreases, environmental degradation increases.
    c. As per capita income increases, environmental degradation decreases.
    d. As per capita income increases, environmental degradation increases.
    e. As per capita income increases, environmental degradation first increases and then decreases.

22. Microlending is the practice of
    a. lending money to businesses to clean up the environment.
    b. loaning small amounts of money to people in less developed countries to start a small business.
    c. being fined for environmental degradation.
    d. lending small sums of money to lower externalities.
    e. borrowing money in order to keep business in the United States rather than abroad.

23. The view that nature has an instrumental value to provide for human needs is
    a. anthropocentric.
    b. biocentric.
    c. ecocentric.
    d. the precautionary principle.
    e. anthropogenic.

24. If environmentally harmful activities are taxed it is a(n)
    a. externality.
    b. cost/benefit evaluation.
    c. green tax.
    d. environmental tax punishment.
    e. cradle to grave tax.

25. When minorities and lower socio-economic families are subjected to higher levels of air pollutants it is a(n)
    a. poverty issue.
    b. environmental justice issue.
    c. microlending issue.
    d. human poverty index issue.
    e. OSHA issue.

ANSWER KEY

Chapter 1
Match These Key Terms/Fill-in-the-Blanks

Environmental Science offers important insights into our world and how we influence it
1. h
2. g
3. f
4. e
5. b
6. c
7. d
8. a

Environmental scientists monitor natural systems for signs of stress
1. b
2. f
3. c
4. a
5. d
6. e
7. i
8. g
9. j
10. h

Human well-being depends on sustainable practices
1. life
2. ecological footprint

Science is a process
1. d
2. n
3. c
4. f
5. m
6. i
7. o
8. e
9. l
10. b
11. a
12. h
13. j
14. g
15. k

Environmental science presents unique challenges
1. equal

After You Read
Short Answer

1. Biology and ecology, toxicology, atmospheric sciences, chemistry, earth sciences, law, literature and writing, ethics, politics and policy, and economics
2. Biodiversity, abundant food production, global surface temperature, the size of the human population and resource depletion
3. Genetic diversity is a measure of the genetic variation among individuals in a population. Species diversity is the number of species in a region or in a particular type of habitat. Ecosystem diversity is a measure of the diversity of ecosystems or habitats that exist in a given region.
4. Anthropogenic activities come from humans. Examples include burning fossil fuels, driving cars, cutting down forests, habitat loss from human construction, etc.
5. 6.8 billion and growing
6. A measure of how much that person consumes, expressed in area of land required to support a person's lifestyle.
7. Observe and question, form testable hypothesis, collect data, interpret results, disseminate findings

Do The Math

50,000 acres= ___20,000___ hectares

75,000 acres= ___30,000___ hectares

150,000 acres= ___60,000___ hectares

Chapter 2
Match These Key Terms

All environmental systems consist of matter

1. m
2. o
3. d
4. i
5. r
6. q
7. dd
8. c
9. k
10. y
11. g
12. cc
13. t
14. a
15. z
16. v
17. e
18. aa
19. h
20. b
21. f
22. ee
23. j
24. p
25. n
26. ff
27. w
28. bb
29. l
30. x
31. u
32. s

Energy is a fundamental component of environmental systems

1. e
2. g
3. i
4. l
5. k
6. m
7. a
8. n
9. d
10. b
11. h
12. f
13. j
14. c

System analysis shows how matter and energy flow in the environment

1. e
2. b
3. d
4. g
5. i
6. f
7. a
8. h
9. c

After You Read
Short Answer

1. Radioactive decay is the spontaneous release of material from the nucleus. Radioactive decay changes the radioactive element into a different element. For example, uranium-235 decays to form thorium-231.

2. An element's half-life is the time it takes for one-half of the original radioactive parent atoms to decay. It is important to know because some elements that undergo radioactive decay emit harmful radiation. Knowing the half-life will allow scientists to determine the length of time an element may be dangerous.

3. The properties of water are surface tension, capillary action, a high boiling point, and the ability to dissolve many different substances.

4. 100 times… because each step up the scale is a factor of 10 times. So, from 3 to 4 would be 10 times more acidic and from 4 to 5 is another 10 times more acidic. 10 X 10 = 100.

5. An example of potential energy is water stored behind a dam. When the water flows downstream, that potential energy becomes kinetic energy.

6. The first law of thermodynamics is that energy cannot be created or destroyed, so the energy never goes away. The second law of thermodynamics is that the energy is transformed and the amount of energy is the same but it cannot do the same amount of work. Most of the energy is converted into heat.

7. Coal burning power plants are approximately 35% efficient, incandescent light bulbs are about 5% efficient and the electrical transmission lines between the power plant and the house is approximately 90% efficient.

8. Mono Lake is an example of a negative feedback loop. When the water level drops there is less lake surface area. With less surface area the evaporation rate decreases. Since there is less evaporation the water in the lake slowly returns to its original volume. The system is responding to a change so it will return to its original state. An example of a positive feedback loop is a population growing. More births create a population increase which in turn creates more births. The amplified population is a positive feedback loop.

Do The Math
1500 watts + 2000 watts= 3500 watts
3500 watts/ 1000watts per kw= 3.5 kw
3.5 kw X .5 (30 mins)= 1.75 kw per 30 mins.
1.75 kw X $.10 = .175 kw per day
.175 kw per day X 7 days per week = $1.23 per week

Review Practice Questions: Chapters 1-2

1. c
2. a
3. c
4. e
5. c
6. d
7. a
8. c
9. b
10. e
11. e
12. a
13. c

14. a
15. b
16. e
17. b
18. e
19. a
20. e
21. d
22. b
23. c
24. e
25. d

Chapter 3
Match These Key Terms/Fill-in-the-Blanks

Ecosystem ecology examines interactions between the living and nonliving world
1. Location

Energy flows through ecosystems
1. d
2. l
3. f
4. c
5. m
6. o
7. i
8. r
9. p
10. e
11. j
12. a
13. h
14. g
15. s
16. b
17. n
18. q
19. k

Matter cycles through the biosphere
1. d
2. g
3. i
4. a
5. j
6. f
7. e
8. h
9. b
10. c

Ecosystems respond to disturbance
1. e
2. b
3. c
4. f
5. a
6. d

Ecosystem provide valuable services
1. c
2. b
3. a

After You Read
Short Answer

1. Solar energy + 6 H2O + 6 CO2 ⟶ C6H12O6 + 6 O2
2. Energy + 6 H2O + 6 CO2 ⟵ C6H12O6 + 6 O2
3.

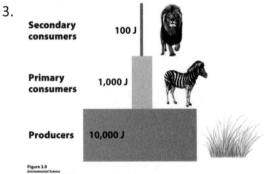

Figure 3.9
Environmental Science
© 2012 W. H. Freeman and Company

4. Fast carbon is carbon that is associated with living organisms and is cycling in the environment. Slow carbon is carbon that is held in rocks, in soils, or as petroleum hydrocarbons. This carbon may be stored for millions of years.

5. Nitrogen fixation produces ammonia. Nitrification produces nitrates and nitrites. Assimilation produces proteins. Ammonification produces ammonia. Denitrification produces nitrogen gas.

Do The Math:
300 trees X $75= <u>$22,500</u>
300 trees X 2 gallons per tree= <u>600 gallons per day</u>
600 gallons per day X 365 days per year= <u>219,000 gallons of water per year</u>
$1.25 per sapling X 300 saplings = <u>$375</u>

Chapter 4
Match These Key Terms

Global processes determine weather and climate
1. f
2. k
3. n
4. a
5. q
6. h
7. g
8. d
9. e
10. m
11. p
12. i
13. b
14. l
15. j
16. o
17. c

Variations in climate determine the dominant plant growth forms of terrestrial biomes
1. c
2. e
3. g
4. i
5. b
6. k
7. j
8. a
9. f
10. h
11. d

Aquatic biomes are categorized by salinity, depth, and water flow
1. c
2. d
3. k
4. g
5. n
6. i
7. m
8. f
9. l
10. a
11. h
12. j
13. e
14. b

After You Read
Short Answer

1. Unequal heating of Earth by the Sun, atmospheric convection currents, the rotation of Earth, Earth's orbit around the Sun on a tilted axis, and ocean currents.

2.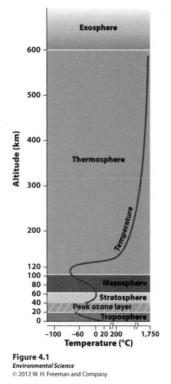

Figure 4.1
Environmental Science
© 2012 W. H. Freeman and Company

3. Coriolis effect, atmospheric convection currents, and the mixing of air currents in the mid-latitudes.

4. There are many ways oceans regulate the temperature of Earth. First, gyres, or large-scale water circulation, redistribute heat in the atmosphere. Also, when surface currents diverge, it causes deeper waters to rise and replace the water that has moved away and this water is generally colder. Thermohaline circulation is another way, which is the sinking of dense, salty water in the North Atlantic moving cold water around the world. Finally, every 3 to 7 years, the surface currents in the tropical Pacific Ocean reverse direction allowing warm water from the equator to move, known as El Niño-Southern Oscillation.

5. Tundra, boreal forest, temperate seasonal forest, woodland/shrubland, temperate grassland/cold desert, temperate rainforest, tropical rainforest, tropical seasonal forest/savanna, subtropical desert.

6. Streams, rivers, lakes, ponds, freshwater wetlands, salt marshes, mangrove swamps, intertidal zone, coral reefs, the open ocean.

Chapter 5
Match These Key Terms

Earth is home to a tremendous diversity of species

1. b
2. d
3. f
4. a
5. e
6. c

Speciation and extinction determine biodiversity

1. d
2. a
3. f
4. e
5. b
6. c

Evolution is the mechanism underlying biodiversity

1. h
2. j
3. b
4. e
5. l
6. g
7. o
8. n
9. a
10. i
11. m
12. d
13. k
14. c
15. f

Evolution shapes ecological niches and determines species distributions

1. d
2. b
3. f
4. a
5. g
6. h
7. c
8. e

After You Read

Short Answer:

1. Species evenness measures the relative proportions of species in an area compared to the total number of species in that area. For example, if a community has 25% of species A, 25% of species B, 25% of species C, and 25% of species D, the community is very even. Species richness refers to the number of different species in a given area. High richness means there are a lot of different species. Low richness means there are few different kinds of species in the area.

2. The key ideas are that individuals produce an excess of offspring, not all offspring can survive, individuals differ in their traits, differences in traits can be passed on from parents to offspring and differences in traits are associated with differences in the ability to survive and reproduce.

3. - Mutations are random changes in the genetic code. This can be a positive or a negative change.
   - Genetic drift is where a small population can lose some of its genetic composition over time as a result of random mating.
   - The bottleneck effect is when there is a drastic reduction in the size of a population and therefore the population's genetic composition is changed.
   - The founder effect is when a few individuals of a population colonize an island and only those that are on the island can now mate. This can change the species on the island very dramatically until they are no longer similar to the population on the mainland.

4. The figure is showing how an original field mouse population has been separated by a river and can no longer breed with one another. This caused two different and genetically distinct populations to form that may not be able to interbreed.

5. The factors that determine the pace of evolution are: if the rate of the environmental change is relatively slow, if the population has high genetic variation for selection to act on, if the population is relatively small, and if the generation time is short.

6. Scientists believe the five causes are habitat destruction, overharvesting, introductions of invasive species, climate change, and emerging diseases.

Review Practice Questions: Chapters 3-5

1. e
2. a
3. d
4. a
5. c
6. d
7. c
8. a
9. e
10. b
11. b
12. e
13. a

14. b
15. d
16. c
17. c
18. e
19. b
20. d
21. b
22. a
23. a
24. b
25. d

Chapter 6

Match These Key Terms/Fill-in-the-Blanks

Nature exists at several levels of complexity

1. Individual
2. Species, area, time
3. Populations, area
4. Ecosystem
5. Biosphere

Population ecologists study the factors that regulate population abundance and distribution

1. d
2. j
3. i
4. g
5. b
6. f
7. c
8. h
9. e
10. a

Growth models help ecologists understand population changes

1. d
2. j
3. a
4. g
5. l
6. h
7. m
8. e
9. i
10. f
11. k
12. b
13. c

Community ecologists study species interactions
1. d
2. h
3. j
4. f
5. b
6. o
7. k
8. e
9. n
10. p
11. g
12. m
13. l
14. a
15. i
16. c

The composition of a community changes over time
1. c
2. a
3. b
4. d

The species richness of a community is influenced by many factors
1. Habitat size, distance

After You Read

Short Answer

1.

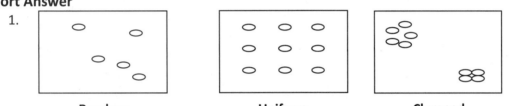

Random   Uniform   Clumped

2. The availability of food is an example of a density-dependent factor. Hurricanes, tornadoes, floods, fires, and volcanic eruptions are examples of density-independent factors.

3.

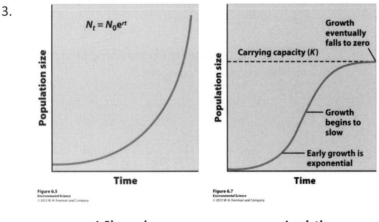

J-Shaped   Logistic

4.

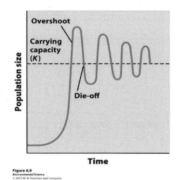

Figure 6.9
Environmental Science
© 2012 W. H. Freeman and Company

As the hares increase in number there is more food for the lynx. As the lynx starts to eat more hares the lynx increase in numbers and the hares decrease in number. This cycle continues.

5. Mutualism is when both species benefit from each other. Commensalism is when one species benefits and another species neither benefits nor is harmed. Parasitism is when one species benefits and the other species is harmed.

6. Primary succession is when a community starts with bare rock devoid of soil. This type of succession takes a very long time. Secondary succession is when an area has been disturbed by something like a fire but the soil is left. Secondary succession is much faster than primary succession.

Do The Math

Year 1:	Year 5:	Year 10:
N= 15 X $e^{.25 \times 1}$	N= 15 X $e^{.25 \times 5}$	N= 15 X $e^{.25 \times 10}$
N= 15 X 1.28	N= 15 X 3.49	N= 15 X 12.18
N= 19 deer	**N= 52 deer**	**N= 182 deer**

Chapter 7
Match These Key Terms/Fill-in-the-Blanks

Many factors drive human population growth
1. g
2. a
3. d
4. i
5. f
6. m
7. k
8. o
9. q
10. c
11. j
12. r
13. n
14. h
15. e
16. p
17. l
18. b

Many nations go through a demographic transition
1. subsistence economy, industrialization, affluence
2. number, spacing

Population size and consumption interact to influence the environment
1. a
2. c
3. d
4. b

After You Read

Short Answer

1. (15+5)- (10+2)/10 = 20-12/10 = 8/10= .8 %
2. Doubling time= 70/.8 = 87.5 years

3.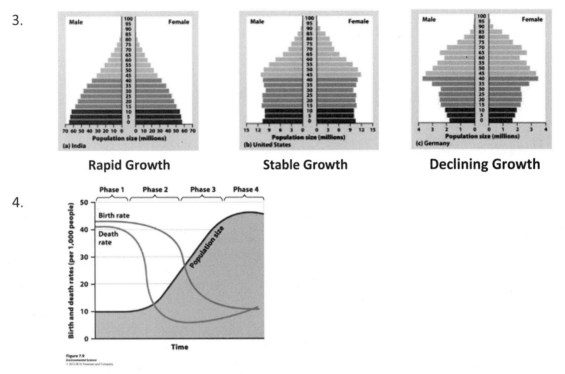

| Rapid Growth | Stable Growth | Declining Growth |

4.

5. IPAT: Impact= Population x Affluence x Technology. In other words, the three major factors that influence environmental impact are how big the population is, how wealthy the population is, and how technologically advanced the population is. The larger the IPAT, the larger the environmental impact.
6. The four types of economic activity that make up GDP are consumer spending, investments, government spending, and exports minus imports.

Review Practice Questions: Chapters 6-7

1. d
2. b
3. c
4. a
5. b
6. e
7. c
8. a
9. e
10. d
11. b
12. a
13. c

14. a
15. e
16. b
17. d
18. a
19. e
20. b
21. d
22. c
23. a
24. d
25. b

Chapter 8
Match These Key Terms

The availability of Earth's resources was determined when the planet formed

1. b
2. f
3. e
4. a
5. d
6. c

Earth is dynamic and constantly changing

1. m
2. e
3. d
4. l
5. a
6. n
7. c
8. o
9. b
10. k
11. j
12. i
13. f
14. h
15. g

The rock cycle recycles scarce minerals and elements

1. b
2. g
3. d
4. e
5. f
6. c
7. i
8. a
9. j
10. k
11. l
12. h

Soils link the rock cycle and the biosphere

1. i
2. h
3. g
4. l
5. c
6. f
7. b
8. d
9. k
10. e
11. a
12. j

The uneven distribution of mineral resources has social and environmental consequences

1. e
2. b
3. f
4. a
5. h
6. j
7. g
8. i
9. d
10. c

After You Read
Short Answer

1.

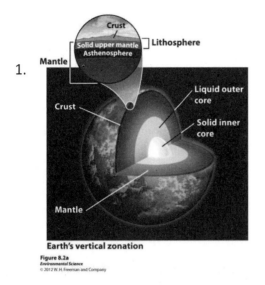

Crust

Lithosphere

Solid upper mantle
Asthenosphere

Mantle

Liquid outer core

Crust

Solid inner core

Mantle

Earth's vertical zonation

Figure 8.2a
Environmental Science
© 2012 W. H. Freeman and Company

2. Plate tectonics state that Earth's lithosphere is divided into plates, most of which are in constant motion. New lithosphere is added at spreading zones and older lithosphere is recycled into the mantle at Subduction zones.

3.

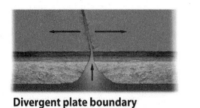  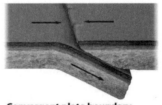 

Divergent plate boundary

Figure 8.8a
Environmental Science

Convergent plate boundary

Figure 8.8b
Environmental Science

Transform fault boundary

Figure 8.8c
Environmental Science

4. Each time you move up the Richter scale you are increasing the damage done by an earthquake by a multiple of **10**.

5. Weathering, erosion, transportation and compression form sedimentary rock.

6.

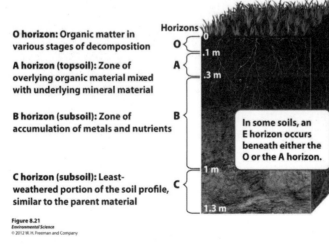

Horizons

O horizon: Organic matter in various stages of decomposition — O — 0, .1 m

A horizon (topsoil): Zone of overlying organic material mixed with underlying mineral material — A — .3 m

B horizon (subsoil): Zone of accumulation of metals and nutrients — B

In some soils, an E horizon occurs beneath either the O or the A horizon.

C horizon (subsoil): Least-weathered portion of the soil profile, similar to the parent material — C — 1 m, 1.3 m

Figure 8.21
Environmental Science
© 2012 W. H. Freeman and Company

7. The five properties that determine soil are parent material, climate, topography, organisms, and time.
8. Figure 8.23 is showing how sand is very porous and therefore water flows through it quickly. Silt is less porous so water flows through it slower than sand. Clay is impermeable and therefore holds water and the water does not move through it.
9. Surface mining is when the mineral or ore you want to remove is close to Earth's surface. Subsurface mining is when the desired resource is more than 100 m below Earth's surface.

Do The Math
500 km= 500,000 m= 500,000,000 mm
500,000,000 mm/15mm = 33,333,333 years

Chapter 9
Match These Key Terms/Fill-in-the-Blank

Water is abundant, but usable water is rare
1. G
2. E
3. M
4. J
5. C
6. N
7. F
8. K
9. L
10. A
11. I
12. H
13. D
14. B

Humans can alter the availability of water
1. F
2. C
3. B
4. G
5. E
6. A
7. D

Water is used by humans for agriculture, industry, and household needs
1. roots, nutrient-rich, soil

After You Read
Short Answer
1. An unconfined aquifer is an aquifer where water can easily flow in and out. A confined aquifer has an impermeable, or confining, layer which impedes water flow to or from the aquifer.
2. Saltwater intrusion occurs when saltwater mixes with freshwater and contaminates well water. It is common in coastal areas. This happens when too many wells are drilled and the water pressure is lowered allowing the saltwater to move into the aquifer. With fewer wells drilled, the water pressure from the freshwater is higher and the saltwater cannot intrude.
3. A body of water that has too much nitrogen or phosphorus can become eutrophic. This means there is a high level of productivity.
4. Freshwater wetlands are important because they help to control flooding, and can absorb and store excess water and release it slowly.

5. The construction of impermeable and paved surfaces (buildings and pavement) by humans has impacted flooding. The surfaces do not allow water to infiltrate into the ground. This causes the water to run off into storm sewers or streams.
6. Benefits of dams include water for human consumption, generation of electricity, flood control and recreation. Consequences of dams include having to displace people, flooding an area that was not flooded, interruption of the water flow that many organisms need, and loss of breeding grounds for animals such as salmon.
7. 70% of water is used for irrigation of crops, 20% for industry and 10% for household uses.

Do The Math
8 loads X 4 weeks= 32 loads per month
200L-100L/load X 32 loads per week= 320 L per month
200L X 32 loads per month = 6400L for the traditional machine
100L X 32 loads per month = 3200L for the front loader machine
6400L – 3200L = 3200L of water saved

$.035/1000L X 320 L/month = $0.11 savings per month
$.035/1000L X 3200L/month= $1.12 savings per month

Review Practice Questions: Chapters 8-9

1. b
2. a
3. c
4. b
5. d
6. e
7. e
8. c
9. e
10. d
11. d
12. d
13. b
14. e
15. b
16. e
17. c
18. a
19. d
20. b
21. a
22. c
23. e
24. d
25. a

Chapter 10
Match These Key Terms/Fill-in-the-Blanks

Human land use affects the environment in many ways
1. c
2. a
3. b

Public lands are classified according to their use
1. maximize, greatest
2. recreation, grazing, timber harvesting, mineral extraction

Land management practices vary according to land use
1. g
2. b
3. m
4. h
5. a
6. j
7. i
8. d
9. k
10. l
11. e
12. c
13. f

Residential land use is expanding
1. i
2. a
3. g
4. e
5. n
6. c
7. m
8. d
9. k
10. h
11. j
12. l
13. b
14. f

After You Read
Short Answer

1. Change to the landscape is the single largest cause of species extinctions.
2. There are many examples, including the one in the textbook: if a bakery moves next door and they begin baking at 3am every morning, making so much noise that they interrupt your sleep and you are not productive at your job the next day. Another example might be neighbors add cows to their property and the resulting fly population limits the enjoyable time you can spend in your yard with the increase of annoying pests.
3. Some of the uses are timber production, defense, urban, residential, transportation, recreational and wildlife lands, cropland, forest grazing land, and grassland/grazing land.
4. When fires are suppressed, the accumulation of large quantities of dead biomass can build up and a large fire can occur that destroys land and property.
5. Urban sprawl is when people begin to move away from urban and rural areas and into suburban and exurban areas. Urban blight is when the population shifts to the suburbs, and therefore the city's revenue begins to shrink. This can lead to declining neighborhoods, businesses leaving the city, city services being cut and city budgets falling.

Chapter 11
Match These Key Terms

Human nutritional requirements are not always satisfied

1. g
2. h
3. c
4. d
5. a
6. f
7. b
8. e

The Green Revolution and industrial farming methods have transformed agriculture

1. k
2. l
3. d
4. e
5. n
6. h
7. m
8. g
9. p
10. r
11. a
12. s
13. c
14. i
15. j
16. b
17. q
18. o
19. f

Alternatives to industrial farming methods are gaining more attention

1. b
2. j
3. f
4. g
5. k
6. e
7. a
8. d
9. c
10. h
11. l
12. i

Modern agribusiness includes the farming of meat and fish

1. a
2. c
3. f
4. d
5. e
6. b

After You Read
Short Answer

1. Reasons for undernutrition and malnutrition include poverty, political and economic factors, political unrest, rise in food prices, and food being diverted to feed livestock and poultry.
2. Most energy subsidies in modern agriculture go to producing fertilizers and pesticides, operating tractors, pumping water for irrigation, and harvesting and preparing food.
3. Irrigation can contribute to soil degradation by being waterlogged (soil remains under water for prolonged periods resulting in impaired root growth due to lack of oxygen) or when small amounts of salts in irrigation water become highly concentrated on the soil surface through evaporation (salinization).

4. Disadvantages of synthetic fertilizers include the use of fossil fuels, runoff of the fertilizer into bodies of water which can reduce oxygen levels, and they do not add organic matter to the soil.
5. Bioaccumulation is when persistent pesticides such as DDT accumulate in fatty tissues of animals and when an animal with DDT in its system is eaten by another animal the chemical is transferred to the consumer. This can lead to very high pesticide concentrations at high trophic levels.
6.

Benefits of GMOs	Concerns about GMOs
Increased crop yield and quality	Worries about human safety
Less pesticide use	Effects on biodiversity
Increased profits	Who will regulate the crops

7. Possible answers include intercropping, crop rotation, agroforestry, contour plowing, and no-till agriculture.
8. There is evidence that antibiotics given to confined animals are contributing to an increase in antibiotic-resistant strains of microorganisms that can affect humans. Another concern is waste disposal of the manure. Runoff into waterways can also be a problem.

Do The Math
1. 2,200kilocalories per day X 30 days per month= 66,000 kilocalories per month
2. 2,200 kilocalories per day / 53 kilocalories per apple= 41.5 apples per day
   66,000 kilocalories per month/53 kilocalories per apple = 1245 apples per month or 41.5 apples per day X 30 days = 1245 apples per month
3. 6,800,000,000 people X 41.5 per day= 2.8×10^{11} apples per day

Review Practice Questions: Chapters 10-11

1. a
2. d
3. d
4. c
5. b
6. e
7. b
8. c
9. a
10. d
11. c
12. e
13. d
14. a
15. a
16. c
17. b
18. e
19. c
20. e
21. d
22. b
23. e
24. c
25. a

Chapter 12
Match These Key Terms

Nonrenewable energy accounts for most of our energy use
1. b
2. d
3. c
4. e
5. a

Fossil fuels are a finite resource
1. b
2. a
3. c

Electricity is a convenient form of energy
1. f
2. a
3. e
4. b
5. c
6. d

Nuclear energy is getting a second look
1. e
2. d
3. b
4. g
5. a
6. c
7. f

Fossil fuels provide most of the world's energy
1. a
2. f
3. e
4. b
5. c
6. d

After You Read
Short Answer

1. Commercial energy sources are bought and sold, such as coal, oil, and natural gas. Subsistence energy sources such as wood are gathered and used by individuals for their own needs.
2. Energy consumption has gone up, particularly the use of coal, oil, nuclear and natural gas. Wood and hydroelectricity have stayed relatively low.
3. The reason is because more passengers can fit on a bus than a single passenger in a car.
4. Coal is burned to transfer energy to water, which becomes steam. This steam is transferred to the blades of a turbine. The shaft in the center of the turbine turns the generator, which generates electricity.
5. Steam used for industrial purposes or to heat buildings can be diverted to turn a turbine first, and create electricity.
6. Vegetation dies and is buried under anaerobic conditions, forming peat. Peat is compressed to form lignite. Further compression yields sub-bituminous and bituminous coal and more pressure and in time forms anthracite.
7. Advantages: coal is plentiful, easy to exploit, we have the technology, and economic costs are low. Disadvantages: coal releases sulfur in the atmosphere, contributes to air pollution, and leaves behind ash.

8. Advantages: petroleum is convenient to transport and use, it is relatively energy-dense and burns cleaner than coal. Disadvantages: petroleum releases sulfur, mercury, lead, and arsenic to the atmosphere when burned, oil spills can occur, and habitat loss can occur.

9. Advantages: natural gas has fewer impurities and therefore emits almost no sulfur dioxide or particulates when burned, and natural gas emits only 60% as much carbon dioxide as coal. Disadvantages: Methane can escape into the atmosphere which is a greenhouse gas, the process of drilling and opening the rock can release the gas, and large quantities of water are used when processing and this can contaminate groundwater.

10. Advantages: nuclear energy does not produce air pollution, and countries with limited fossil fuel resources can obtain energy independence. Disadvantages: possibility of accidents and disposal of radioactive waste.

Do The Math

½= 700,000,000 million years

¼= 1,400,000,000 million years

1/8= 2,800,000,000 million years

Chapter 13
Match These Key Terms/Fill-in-the-Blanks

What is renewable energy?
1. c
2. a
3. b
4. d

How can we use less energy
1. d
2. e
3. a
4. c
5. b

Biomass is energy from the Sun
1. d
2. a
3. e
4. f
5. b
6. g
7. c

The kinetic energy of water can generate electricity
1. c
2. b
3. d
4. a
5. e

The Sun's energy can be captured directly
1. energy of sunlight
2. light, heat, electricity

Earth's internal heat is a source of nondepletable energy
1. radioactive decay, Earth
2. high, inertia

Wind energy is the most rapidly growing source of electricity
1. kinetic energy
2. electricity

Hydrogen fuel cells have many potential applications
1. common battery
2. electric current, split, oxygen

How can we plan our energy future?
1. electricity, electricity, end

After You Read

Short Answer

1. Three possible answers could be to lower thermostat during cold months, drive fewer miles, and turn off the computer when not used.

2. Some ways to utilize passive solar design in the northern hemisphere would be to construct a house with windows on the south-facing wall, use double-paned windows, place windows so they get the most light into the building, use dark materials on the roof to absorb solar energy, build roof overhangs, and/or use window shades.

3. Modern carbon is carbon that is in plants growing today and in the atmosphere today. Fossil carbon is carbon that has been buried for millions of years.

4. Some answers are: corn, corn by-products, wood chips, sugar cane, crop waste, switchgrass, algae, soybeans, and palms.

5.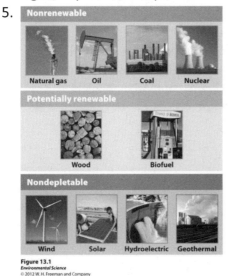

Figure 13.1
Environmental Science
© 2012 W. H. Freeman and Company

Do The Math

$0.10/hour X 10 hours per day = $1.00 savings per day
$200 savings would take 200 days.

Review Practice Questions: Chapters 12-13

1. e
2. b
3. b
4. d
5. c
6. a
7. a
8. d
9. a
10. e
11. d
12. b
13. c
14. e
15. d
16. a
17. c
18. e
19. a
20. b
21. b
22. e
23. c
24. d
25. d

Chapter 14
Match These Key Terms/Fill-in-the-Blanks

Pollution can come from specific sites or broad areas
1. contamination, human activities, organisms
2. distinct, factory, discharges
3. Nonpoint sources, suburban community, runoff

Human wastewater is a common pollutant
1. c
2. g
3. d
4. b
5. f
6. a
7. e

We have technologies to treat wastewater from humans and livestock
1. f
2. d
3. c
4. e
5. b
6. a

Heavy metals and other substances can pose serious threats to human health and the environment
1. Earth, rain, snow, particles
2. PCBs, compounds, environmental
3. concern, uneasiness

Not all water pollutants are chemicals
1. human activities, temperature, water
2. water, thermal shock

A nation's water quality is a reflection of the nation's water laws and their enforcement
1. Safe, EPA, maximum contaminant levels, surface, groundwater

After You Read
Short Answer

1. Point source pollutants can come from a distinct location such as a factory or sewage treatment plant. Nonpoint source pollutants come from a large area such as an entire farming region, a suburban community with many lawns and septic systems, or storm runoff from parking lots.
2. Wastewater dumped into bodies of water naturally undergoes decomposition by bacteria, which creates a large demand for oxygen in the water. Second, the nutrients that are released from wastewater decomposition can make the water more fertile, and third, wastewater can carry a wide variety of disease-causing agents.

3.

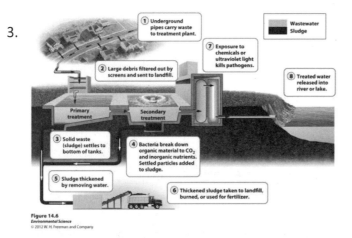

Figure 14.6
Environmental Science
© 2012 W. H. Freeman and Company

4. Lead can come from pipes of older homes, brass fittings containing lead, and solder. Arsenic is found naturally in the Earth's crust but can get into groundwater from mining and industrial uses such as wood preservatives. Mercury comes from the burning of fossil fuels, incineration of garbage, hazardous waste, medical supplies, and dental supplies.
5. DDT was designed to target insects, but moved up the food chain all the way to eagles that consume fish. Eagles produced eggs with thinner shells that would break during incubation.
6. Thermal pollution can kill organisms that are not adapted to the higher temperatures. It can also cause organisms to increase their respiration rates, and warmer water does not contain as much dissolved oxygen as cold water.

Do The Math
1. 60 L of manure a day X 460 animals = 27,600 L a day.
2. 60 L of manure a day X 460 animals X 7 days/week = 193,200 L a week.
3. 60 L of manure a day X 460 animals X 365 days/year = 10,074,000 L a year.

Chapter 15
Match These Key Terms/Fill-in-the-Blanks

Air pollutants are found throughout the entire global system
1. i
2. e
3. f
4. c
5. b
6. g
7. a
8. h
9. d

Photochemical smog is still an environmental problem in the United States
1. warm, cold, dense
2. warm, emissions

The stratospheric ozone layer provides protection from ultraviolet solar radiation
1. chlorine, anthropogenic

Indoor air pollution is a significant hazard, particularly in developing countries
1. insulating, respiratory diseases, asbestosis, lung cancer
2. toxic, airtight

After You Read

Short Answer

1.

Compound	Effects
Sulfur dioxide	Respiratory irritant, damage plant tissue, harmful to aquatic life
Nitrogen dioxide	Respiratory irritant, forms photochemical smog, harmful to aquatic life, over fertilizes land and water
Carbon monoxide	Interferes with oxygen transport, headaches, can cause death
Particulate matter	Respiratory and cardiovascular disease, reduced lung function, contributes to haze and smog
Lead	Impairs central nervous system, effects learning and concentration
Ozone	Reduces lung function, degrades plant surfaces, damages rubber and plastic
VOCs	Precursor to ozone formation
Mercury	Impairs central nervous system, bioaccumulates up the food chain
Carbon dioxide	Affects climate, greenhouse gas

2. Photochemical smog is dominated by oxidants such as ozone where sulfurous smog is dominated by sulfur dioxide and sulfate compounds.

3. Anthropogenic causes of air pollution are transportation, electricity generation, natural and human-made fires, and road dust.

4. A thermal inversion is when a warm layer traps emissions that then accumulate beneath the inversion layer. These create pollution events and are particularly common in some cities, where high concentrations of vehicle exhaust and industrial emissions are easily trapped by the inversion layer.

5. The effects of acid deposition include lowering the pH of lake water, decreased species diversity of aquatic organisms, mobilizing metals in soils and surface water, changes in trophic levels, damage to plants, and damage to human-built structures like statues, monuments, and buildings.

6. In Mexico City they plan to permit automobiles to only be driven every other day. In China, during the Olympics, they expanded public transportation and imposed motor vehicle restrictions, temporarily shutting down a number of industries. Some cities in England are experimenting with charging individual user fees to use roads at certain times of the day.

7. UV-C radiation breaks the bonds holding together the oxygen molecule, leaving two free oxygen atoms. The result is that in the presence of UV radiation, oxygen is converted to ozone. Ozone is broken down into O2 and free oxygen atoms. The free oxygen atoms and molecular oxygen may again react to produce ozone.

8. CFCs were used in refrigeration, air conditioners, propellants in aerosol cans, and in foam products like Styrofoam cups and foam insulation.
9. Some indoor air pollutants are VOCs, asbestos, carbon monoxide, radon, tobacco smoke, paints and cleaning fluids.

Do The Math

1. 384 ppm-376 ppm/ 376ppm X 100%= 2%

Chapter 16

Match These Key Terms/Fill-in-the-Blanks

Humans generate waste that other organisms cannot use

1. c
2. b
3. a

The three Rs and composting divert materials from the waste stream

1. c
2. d
3. e
4. g
5. b
6. f
7. a

Currently, most solid waste is buried in landfills or incinerated

1. d
2. f
3. h
4. g
5. e
6. a
7. b
8. c

Hazardous waste requires special means of disposal

1. b
2. a
3. c

There are newer ways of thinking about solid waste

1. tool, used, released
2. waste reduction, disposal strategies, impact

After You Read

Short Answer

1. Organic material such as wood, yard trimmings, food scraps, and paper make up 64% of MSW, glass makes up 5%, metals 8%, plastics 12%, rubber, leather, and textiles 8%, and other 3%.
2. Parts of a landfill include leachate collection systems, a liner with sand and clay, and a method to monitor the groundwater for any leaks.
3. Leachate found in landfills might contaminate underlying and adjacent waterways, landfills can pose a risk to human health, and the methane buildup could lead to explosions.
4. Problems with waste incineration include: ash that is left is more concentrated and more toxic, the cost of the practice, and air pollution.
5. RCRA (Resource Conservation and Recovery Act) was created to protect human health and the natural environment by reducing or eliminating the generation of hazardous waste. RCRA is known as "cradle to grave" tracking.

CERCLA (Comprehensive Environmental Response, Compensation, and Liability Act) is also known as Superfund and authorizes the federal government to respond directly to the release or threatened release of substances that may pose a threat to human health or the environment.

6. Integrated Waste Management is using several waste reduction, management, and disposal strategies in order to reduce the environmental impact of MSW. These consist mainly of the 3 R's: reduce, reuse, and recycle.

Do The Math

1. 250 mm/year = 0.250m/year

   0.250m/year X 10,000m^2 X 50%= 1250 m^3

2. 1250m^3 X 80%= 1000 m^3

Chapter 17
Match These Key Terms

Human health is affected by a large number of risk factors
1. c
2. d
3. b
4. a

Infectious disease have killed large numbers of people
1. j
2. d
3. c
4. g
5. b
6. i
7. m
8. k
9. h
10. f
11. e
12. l
13. a

Risk analysis helps us assess, accept, and manage risk
1. e
2. d
3. b
4. c
5. a

Toxicology is the study of chemical risks
1. e
2. b
3. c
4. f
5. d
6. a

Scientists can determine the concentrations of chemicals that harm organisms
1. f
2. d
3. h
4. l
5. g
6. k
7. i
8. c
9. m
10. a
11. n
12. b
13. e
14. j

After You Read

Short Answer

1. Developing nations' risk factors include poverty, unsafe drinking water, poor sanitation, and malnutrition. Developed nations' risks can include tobacco use, less active lifestyles, poor nutrition, and overeating that leads to high blood pressure and obesity.

2.

Disease	How Disease is Spread
Plague	Fleas on rats
Malaria	Mosquitoes
Tuberculosis	Person to person
HIV/AIDS	Person to person
Ebola hemorrhagic fever	Unknown
Mad cow disease	People that ate meat from contaminated cattle
Bird flu	Birds to people
West Nile virus	Mosquitoes

3. Some different pathways include domesticated animals, other humans, water, air, food, and wild animals.

4.

TABLE 17.1	Some chemicals of major concern		
Chemical	**Sources**	**Type**	**Effects**
Lead	Paint, gasoline	Neurotoxin	Impaired learning, nervous system disorders, death
Mercury	Coal burning, fish consumption	Neurotoxin	Damaged brain, kidneys, liver, and immune system
Arsenic	Mining, groundwater	Carcinogen	Cancer
Asbestos	Building materials	Carcinogen	Impaired breathing, lung cancer
Polychlorinated biphenyls (PCBs)	Industry	Carcinogen	Cancer, impaired learning, liver damage
Radon	Soil, water	Carcinogen	Lung cancer
Vinyl chloride	Industry, water from vinyl chloride pipes	Carcinogen	Cancer
Alcohol	Alcoholic beverages	Teratogen	Fetuses with reduced fetal growth, brain and nervous system damage
Atrazine	Herbicide	Endocrine disruptor	Feminization of males, low sperm counts
DDT	Insecticide	Endocrine disruptor	Feminization of males, thin eggshells of birds
Phthalates	Plastics, cosmetics	Endocrine disruptor	Feminization of males

Table 17.1
Environmental Science
© 2012 W. H. Freeman and Company

5. LD50 is how much of a specific toxin it takes to kill 50% of individuals. ED50 is how much of the toxin it takes to have 50% of the individuals experience harmful effects, but not death.

6. DDT is a persistent chemical that increases in concentration as it moves up the food chain. As DDT moved up the food chain it was stored in the fat of the animals that ate it and eventually made its way up to the predatory birds. These birds had eggs that were thin-shelled and often broke when the parent bird incubated the eggs.

7.

TABLE 17.2	The persistence of various chemicals in the environment, measured in terms of their half-life
Chemical	**Half-life**
Malathion insecticide	1 day
Radon	4 days in air
Vinyl chloride	4.5 days in air
Phthalates	4.5 days in water
Roundup herbicide	7 to 70 days in water
Atrazine herbicide	224 days in wetland soils
Polychlorinated biphenyls (PCBs)	8 to 15 years in water
DDT	30 years in soil

Source: Hazardous Substances Data Bank, http://toxnet.nlm.nih.gov/cgi-bin/sis/htmlgen?HSDB/.

Table 17.2
Environmental Science
© 2012 W. H. Freeman and Company

Review Practice Questions: Chapters 14-17

1. d
2. a
3. e
4. b
5. c
6. a
7. b
8. d
9. d
10. c
11. a
12. e
13. b
14. a
15. a
16. e
17. c
18. d
19. e
20. b
21. a
22. c
23. b
24. e
25. d

Chapter 18
Match These Key Terms/Fill-in-the-Blanks

We are in the midst of a sixth mass extinction
1. a
2. c
3. b

Declining biodiversity has many causes
1. b
2. d
3. a
4. f
5. c
6. e

The conservation of biodiversity often focuses on single species
1. international, biodiversity

The conservation of biodiversity sometimes focuses on protecting entire ecosystems
1. two, abrupt, field, forest
2. protected, zones, human

After You Read

Short Answer

1. The five categories used by the International Union for Conservation of Natures are data deficient, extinct, threatened, near-threatened, and least concern.

2. Invasive species can have no natural enemies to control their population.

3. Lacey Act—the earliest law to control trade of wildlife.
   CITES—controls the international trade of threatened plants and animals.
   Marine Mammal Protection Act—prohibits the killing of all marine mammals in the U.S. and prohibits the import or export of any marine mammal body part.
   Endangered Species Act—prohibits the harming of any threatened or endangered species including trade of the species. It also gave the government the right to purchase habitats that are critical to the survival of these species.
   Convention on Biological Diversity—an international treaty to help protect biodiversity.

4. Biosphere reserves consist of core areas that have minimal human impact and outer zones that have increasing levels of human impacts.

Chapter 19

Match These Key Terms/Fill-in-the-Blanks

Global change includes global climate change and global warming

1. A
2. C
3. B

The Kyoto Protocol addresses climate change at the international level

1. Kyoto Protocol
2. carbon dioxide

Solar radiation and greenhouse gases make our planet warm

1. gas, molecule, compound, 100, carbon dioxide

After You Read

Short Answer

1.

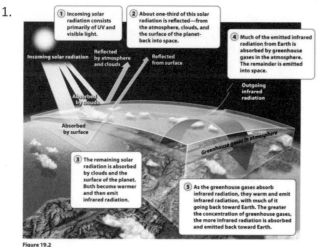

Figure 19.2
Environmental Science
© 2012 W. H. Freeman and Company

2.

TABLE 19.1	The major greenhouse gases		
The major greenhouse gases differ in their ability to absorb infrared radiation and the duration of time that they stay in the atmosphere. The units "ppm" are parts per million.			
Greenhouse gas	Concentration in 2010	Global warming potential (over 100 years)	Duration in the atmosphere
Water vapor	Variable with temperature	<1	9 days
Carbon dioxide	390 ppm	1	Highly variable (ranging from years to hundreds of years)
Methane	1.8 ppm	25	12 years
Nitrous oxide	0.3 ppm	300	114 years
Chlorofluorocarbons	0.9 ppm	1,600 to 13,000	55 to >500 years

Source: Data on concentration are from the National Oceanic and Atmospheric Administration. www.esrl.noaa.gov/gmd/aggi.
Data on global warming potential are from the United Nations Framework Convention on Climate Change.

Table 19.1
Environmental Science
© 2012 W. H. Freeman and Company

3. In the spring, deciduous trees, grasslands, and farmlands in the Northern Hemisphere turn green, so they increase their carbon dioxide intake for photosynthesis. In the winter, they lose their leaves and do not take in as much carbon dioxide.

4. When scientists melt pieces of ice they extract from drilling tubes of ice, they analyze the air bubbles that are released by measuring the concentration of greenhouse gases. They can then determine temperature changes over long periods of time.

5. The two explanations are an increase in solar radiation and the possibility that warming is caused by increased carbon dioxide levels in addition to normal fluctuations.

6. Carbon sequestration is an approach to taking carbon dioxide out of the atmosphere. This might include storing carbon in agricultural soils or retiring agricultural land and allowing it to become pasture or forest. Researchers are also trying to find ways of capturing carbon dioxide from coal-burning power stations.

Do The Math
1. 1.4ppm X 40 years= 56ppm
   390 + 56= 446 ppm in 2050
2. 1.9 X 40 years = 76ppm
   390+76 ppm= 466 ppm in 2050

Chapter 20
Match These Key Terms/Fill-in-the-Blanks

Sustainability is the ultimate goal of sound environmental science and policy
1. status, healthy, prosperous

Economics studies how scare resources are allocated
1. e
2. a
3. d
4. b
5. c

Economic health depends on the availability of natural capital and basic human welfare
1. c
2. d
3. g
4. a
5. b
6. e
7. f

Agencies, laws, and regulations are designed to protect our natural and human capital

1. c
2. g
3. b
4. d
5. i
6. h
7. k
8. j
9. e
10. m
11. f
12. l
13. a

There are several approaches to measuring and achieving sustainability

1. f
2. d
3. e
4. a
5. c
6. b

After You Read

Short Answer

1. Supply is the number of units of an item a manufacturer will provide. Demand is the amount of units of an item consumers want.
2. A microloan is a small loan made to an individual in a developing country to buy items to help them earn an income.
3. The goals are to eradicate extreme poverty and hunger, achieve universal primary education, promote gender equality and empower women, reduce child mortality, improve maternal health, combat HIV/AIDS, malaria, and other diseases, ensure environmental sustainability, and develop a global partnership for development.
4. Dr. Wangari Maathai founded the Green Belt Movement, a Kenyan and international environmental organization that empowers women by paying them to plant trees, some of which can be harvested for firewood in a few years.

Review Practice Questions: Chapters 18-20

1. b	14. d
2. e	15. a
3. a	16. e
4. d	17. b
5. d	18. c
6. b	19. d
7. c	20. a
8. a	21. e
9. c	22. b
10. e	23. a
11. b	24. c
12. c	25. b
13. b	

Full-Length Practice Exam #1

This full-length practice exam contains two parts. Part I consists of 100 multiple-choice questions and Part II consists of four free-response questions.

You will have 90 minutes to complete the multiple-choice section of the exam. This section counts for 60% of the exam grade. As you will not be penalized for incorrect answers, you should answer every question on the test. If you do not know an answer to a question, try to eliminate any incorrect answer choices and take your best guess. Do not spend too much time on any one question. If you know the question is going to take a while to solve, you should skip it and come back to it at the end.

You will have 90 minutes to complete the free-response section of the exam. This section counts for 40% of the overall exam grade. Be sure to answer each part of the question and to provide thorough explanations using the terms and themes you have learned in the course. Also be sure to show your work whenever you use math to solve a problem.

Calculators are not allowed on any portion of the exam.

SECTION I: Multiple Choice

1. Which of the following is true of a country going through the demographic transition and becoming more industrialized?
   - (A) Birth rates rise
   - (B) Death rates rise
   - (C) Birth rates fall
   - (D) Death rates remain high
   - (E) Death rates rise as birth rates fall

2. Which of the following is not associated with passive solar energy?
   - (A) Putting a photovoltaic cell on the roof
   - (B) Putting blinds on a window
   - (C) Planting a deciduous tree outside a west facing window
   - (D) Building a house with a living roof on top
   - (E) Adding extra insulation in the walls and attic of a house

3. Which of the following would be the correct order of soil particles from largest to smallest?
   - (A) Sand, clay, silt
   - (B) Silt, clay, sand
   - (C) Sand, silt, clay
   - (D) Clay, silt, sand
   - (E) Silt, sand, clay

4. A scientist does an experiment with brine shrimp and a particular pesticide and finds that 10 ml of this pesticide kills half of the shrimp. What has the scientist found?
   - (A) The LD-50
   - (B) The effective dose
   - (C) The toxicity level
   - (D) The ED-50
   - (E) The induction zone

5. Which of the following is an example of an indicator species?
   - (A) Elephant
   - (B) Oak tree
   - (C) Bald Eagle
   - (D) Leopard frog
   - (E) Box turtle

6. Which of the following is the best example of the tragedy of the commons?
    (A) A lady with ten pet cats
    (B) A farmer slashing and burning the rainforest
    (C) A person buying up large tracts of land
    (D) A person leaving home with the air conditioner on and the windows open
    (E) Cars and factories making air pollution

7. Radioactive waste in the United States is currently being stored where?
    (A) Yucca Mountain
    (B) In the ocean
    (C) In Mexico
    (D) At nuclear plants where the waste is produced
    (E) At hospitals and other medical facilities

8. What is one concern environmentalists have with people that are overfertilizing their yards?
    (A) Pests becoming resistant to pesticides
    (B) Eutrophication
    (C) Genetically Modified Crops (GMO's)
    (D) Abiotic inclusion
    (E) Drastic pH drop in lakes

9. Approximately 50% of all coal reserves are found in which countries?
    (A) China, Japan, India
    (B) United States, Canada, Mexico
    (C) China, United States, Russia
    (D) Russia, China, Mexico
    (E) India, Brazil, Canada

10. All of the following are chemicals responsible for climate change EXCEPT
    (A) Carbon dioxide
    (B) Methane
    (C) CFCs
    (D) Sulfer Dioxide
    (E) Water vapor

11. Electrostatic precipitators and wet scrubbers are used to remove particulates and SO_2 from
    (A) cars.
    (B) power plants.
    (C) sewage treatment plants.
    (D) landfills.
    (E) contaminated soil.

12. Alum is used in sewage treatments plants as a flocculating reagent. This means that alum helps to
    (A) screen out the sewage.
    (B) decontaminate the sewage.
    (C) get rid of aerobic microorganisms.
    (D) cause the sewage to clump and sink.
    (E) aerate the sewage.

13. The law that states how trash must be picked up, transported, buried, and monitored is the
    (A) RCRA.
    (B) CERCLA.
    (C) FIFRA.
    (D) ESA.
    (E) MMPA.

14. As CFCs enter the stratosphere, they break down ozone molecules. Why is this a concern?
    (A) Ozone is an important component in oxygen.
    (B) Ozone helps to keep our planet at a stable temperature.
    (C) Ozone is a respiratory irritant in the stratosphere.
    (D) Ozone protects us from damaging UV light.
    (E) Ozone helps to get rid of all the carbon dioxide in our atmosphere.

15. The replacement level fertility in some countries is as high as four kids per family. This is because
    (A) women are working outside the home.
    (B) life span is only 75 years.
    (C) of a high infant mortality rate.
    (D) children are not seen as a necessity.
    (E) of a larger population of the elderly.

16. The efficiency of a typical coal burning power plant is
    (A) 100%
    (B) 75%
    (C) 50%
    (D) 30%
    (E) 10%

17. Habitat loss, invasive species, pollution, overpopulation, climate change, and overharvesting are the main reasons that are causing the problem of
    (A) increasing number of endangered species.
    (B) mountain top removal of coal.
    (C) cultural eutrophication in our water ways.
    (D) ozone depletion.
    (E) phosphorus becoming atmospheric.

18. The amount of arable land on earth is
    (A) decreasing.
    (B) doubling.
    (C) being flooded.
    (D) turned into parking lots.
    (E) being turned into rainforest.

19. If a city of population 10,000 has 50 births, 30 deaths, 30 immigrants and 10 emigrants, what is the net annual percentage growth rate?
    (A) 120%
    (B) 40%
    (C) 8%
    (D) .4%
    (E) 10%

20. The Earth is tilted approximately 23 degrees. This causes
    (A) the seasons.
    (B) the tides.
    (C) the solar lights.
    (D) the Coriolis effect.
    (E) hurricanes.

21. If the current human population is approximately 6.8 billion and is growing at an annual rate of 1.17%, approximately how many people will be added next year if this rate stays the same?

   (A) 8×10^5
   (B) 8×10^6
   (C) 8×10^7
   (D) 8×10^8
   (E) 8×10^9

22. Radon is found in bedrock and can seep into homes via basements and other means. Why is radon a concern?

   (A) Radon causes asthma.
   (B) Radon causes brain damage in children.
   (C) Radon damages wood, similar to termites.
   (D) Radon causes lung cancer.
   (E) Radon causes cataracts in humans and animals.

23. All of the following are concerns of nuclear power EXCEPT

   (A) terrorist risk.
   (B) storage.
   (C) safety concerns.
   (D) air pollution.
   (E) radiation.

24. Genetically Modified Crops are grown for all of the following reasons EXCEPT

   (A) they are bug resistant.
   (B) they need less water.
   (C) they are more nutritious.
   (D) they can be grown in places that crops don't normally grow in.
   (E) they use more fertilizers.

25. Which of the following energy sources is nonrenewable?

   (A) Nuclear
   (B) Solar
   (C) Wind
   (D) Geothermal
   (E) Biomass

26. All of the following are considered in the price of a good or service EXCEPT
    (A) material cost.
    (B) labor cost.
    (C) environmental cost.
    (D) transportation cost.
    (E) profit.

27. All weather on Earth is found in which layer of the atmosphere?
    (A) Troposphere
    (B) Stratosphere
    (C) Mesosphere
    (D) Thermosphere
    (E) Exosphere

Use the following graph to answer questions 28 & 29.

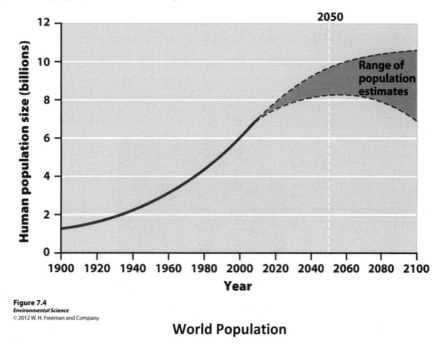

Figure 7.4
Environmental Science
© 2012 W. H. Freeman and Company

World Population

28. According to the graph, from 1800 to 2010 the human population has been growing
    (A) linearly.
    (B) logistically.
    (C) exponentially.
    (D) inversely.
    (E) quadratic.

29. If the size of the human population continues to grow at the high estimate, approximately how many people will live on Earth in the year 2300?

    (A) 11 billion
    (B) 8 billion
    (C) 6 billion
    (D) 20 billion
    (E) 14 billion

30. There are two different communities each with 100 organisms.
    - Community 1 has 10 different species with 10 of each species.
    - Community 2 has 5 different species with the following numbers: species 1 has 40, species 2 has 20, species 3 has 20, species 4 has 10, and species 5 has 10.

    What can we tell about community 1 compared to community 2?

    (A) Community 1 has more richness than community 2
    (B) Community 2 has more richness than community 1
    (C) Community 2 is more even than community 1
    (D) Community 1 has been more disturbed than community 2
    (E) Community 1 is more stable than community 2

Use the following soil horizon to answer questions 31 and 32.

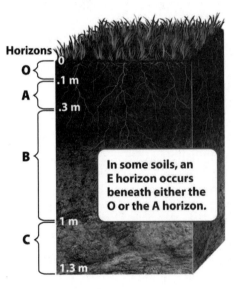

31. Which layer is not true soil, but rather is a layer of organic detritus?

    (A) O
    (B) A
    (C) B
    (D) C
    (E) Both C and D

32. Which layer is known as the subsoil and has very little organic matter?

    (A) O

    (B) A

    (C) B

    (D) C

    (E) Both C and D

33. If a population is growing at a rate of 7% per year, how many years will it take for the population to double?

    (A) 5 years

    (B) 10 years

    (C) 17 years

    (D) 25 years

    (E) 70 years

34. Which season will the level of ozone in the Northern Hemisphere be at its lowest?

    (A) January

    (B) March

    (C) July

    (D) September

    (E) December

35. Where is the world's largest hydroelectric dam?

    (A) United States

    (B) Brazil

    (C) Canada

    (D) China

    (E) India

36. What is the main chemical that causes cultural eutrophication?

    (A) Carbon

    (B) Sulfur

    (C) Oxygen

    (D) Nitrogen

    (E) Calcium

37. A scientist discovers fecal coliform in a river. What would be a good hypothesis of the cause?

    (A) Chemical fertilizers added to yards

    (B) Release of an invasive species to the river

    (C) An improperly working sewage treatment plant

    (D) A factory that is dumping affluent into the river

    (E) Air pollution surrounding the river

38. What is the population density of a China that has 9.6 million km^2 of land area and 1,331 billion people?

    (A) 138.6

    (B) .007

    (C) 12,777

    (D) 1321.4

    (E) 1,331

39. The process of using plants to remove toxic soil is known as

    (A) transpiration.

    (B) eutrophication.

    (C) radiation.

    (D) toxication.

    (E) phytoremediation.

40. Climate change occurs because

    (A) too much UV light is reaching the troposphere.

    (B) IR radiation is trapped in the troposphere.

    (C) too much UV light is reaching the stratosphere.

    (D) IR radiation is trapped in the stratosphere.

    (E) the troposphere is releasing too much IR radiation back into space.

41. A niche specialist would be

    (A) very unlikely to become extinct.

    (B) an animal that eats almost anything.

    (C) a cockroach.

    (D) susceptible to a low pH.

    (E) vulnerable to environmental change.

42. The main reason the human population has been growing exponentially is
    (A) modern medicine.
    (B) Genetically Modified Crops.
    (C) access to reliable birth control.
    (D) Zero Population Growth.
    (E) immigration.

43. Erosion can be caused by all of the following EXCEPT
    (A) sustainable agriculture.
    (B) overgrazing.
    (C) construction projects.
    (D) logging.
    (E) deforestation.

44. What was the approximate population of the United States in 2010?
    (A) 300 million
    (B) 300 billion
    (C) 6.7 million
    (D) 6.7 billion
    (E) 1 billion

45. The order of coal from the lowest energy content to highest energy content is:
    (A) anthracite, lignite, sub-bituminous, bituminous.
    (B) anthracite, sub-bituminous, bituminous, lignite.
    (C) sub-bituminous, bituminous, lignite, anthracite.
    (D) lignite, sub-bituminous, bituminous, anthracite.
    (E) sub-bituminous, bituminous, anthracite, lignite.

46. The fossil fuel that the world has the largest supply of is
    (A) oil.
    (B) natural gas.
    (C) nuclear.
    (D) biomass.
    (E) coal.

Match the following agreements in numbers 47-50 with their primary goal.

47. _____ Montreal Protocol

48. _____ Kyoto Protocol

49. _____ Millennium Development

50. _____ 1994 Global Summit

(A) eradicate poverty

(B) lower climate change

(C) lower ozone depletion

(D) provide access to family planning

(E) stop point source pollution

51. One of the concerns of CAFOs, Concentrated Animal Feeding Operations is
    (A) the amount of air pollution they create.
    (B) the amount of land they use.
    (C) the amount of animal waste runoff they produce.
    (D) SARS (Severe Acute Respiratory Syndrome).
    (E) increase in wind erosion.

52. No-till farming leads to a(n)
    (A) decrease in water erosion.
    (B) increase in wind erosion.
    (C) increase in water erosion.
    (D) overfertilization.
    (E) an increase in microorganisms.

53. What heavy metal is produced by the burning of coal?
    (A) Lead
    (B) Mercury
    (C) Arsenic
    (D) Iron
    (E) Titanium

54. Which type of irrigation is 95% efficient?
    (A) Spray irrigation
    (B) Drip irrigation
    (C) Flood irrigation
    (D) Pivot irrigation
    (E) Furrow irrigation

55. Economic activity that creates a cost for a third party that is not borne by either the buyer or seller is
    (A) a negative externality.
    (B) a positive externality.
    (C) an internalized cost.
    (D) eutrophication.
    (E) a fixed cost.

56. If the production of a good or service creates a negative externality, then
    (A) the price is too high and too much is produced.
    (B) the price is too high and too little is produced.
    (C) the price is too low and too little is produced.
    (D) the price is too low and too much is produced.
    (E) the market price accurately reflects the true cost of production.

57. If your air conditioner uses 600 watts of energy per hour on a daily basis, and your energy cost is $.10 per kWh, and there are 30 days in the month, how much does the energy used by the air conditioner cost you per month?
    (A) $1.30
    (B) $4.32
    (C) $43.20
    (D) $13.01
    (E) $1.80

58. If approximately 1.1 billion people in the world do not have access to clean water, what percentage of the world does not have clean water?
    (A) 10%
    (B) 15%
    (C) 25%
    (D) 40%
    (E) 50%

59. If you have 1,500 cows, and each cow produces 50 liters of manure a day, how many liters of manure would be produced in 45 days?
    (A) 33kL
    (B) 337kL
    (C) 3,375kL
    (D) 33,750kL
    (E) 337,500kL

60. Which layer of the atmosphere is most dense?
   (A) Troposphere
   (B) Stratosphere
   (C) Mesosphere
   (D) Thermosphere
   (E) Exosphere

61. Which biome has permafrost?
   (A) Tundra
   (B) Grassland
   (C) Tropical rainforest
   (D) Temperate rainforest
   (E) Desert

62. Which scientist is credited with the theory of evolution by natural selection?
   (A) Watson
   (B) Crick
   (C) Mendeleev
   (D) Muir
   (E) Darwin

63. After years of adding a pesticide to his land, a farmer notices that less and less insects are being killed. This is probably because
   (A) the insects have become resistant.
   (B) there are less available nutrients.
   (C) there is less available water.
   (D) of overharvesting of the natural resources.
   (E) of eutrophication in local waterways.

64. What are the negative environmental problems associated with using biomass for energy?
   I. Deforestation
   II. Soil erosion
   III. Air pollution
   (A) I only
   (B) II only
   (C) III only
   (D) I and II
   (E) I, II, and III

Match the energy source in numbers 65-69 with the characteristic that describes it.

65. _____ Solar

66. _____ Wind

67. _____ Geothermal

68. _____ Biomass

69. _____ Hydroelectric

(A) does not come from the sun

(B) is used to cook with in many developing nations

(C) is the most promising of all renewable sources

(D) uses photovoltaic cells

(E) uses falling water to make energy

Use the food chain below to answer question 70.

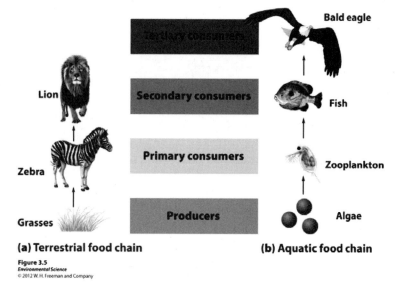

(a) Terrestrial food chain

(b) Aquatic food chain

Figure 3.5
Environmental Science
© 2012 W. H. Freeman and Company

70. If there are 10,000 kilocalories of energy available at the producer level, how many kilocalories will be available at the tertiary level?

(A) 10,000

(B) 1,000

(C) 100

(D) 10

(E) 9

71. "Solar energy + $6H_2O$ + 6 CO_2 → C_6H12O_6 + $6O_2$" is the formula for

(A) respiration.

(B) photosynthesis.

(C) chemosynthesis.

(D) cellular respiration.

(E) primary productivity.

72. A forest ecosystem has an NPP of 3.05 kg C/m2/year and a GPP of 4.5 kg C/m2/year. How much carbon is being used during respiration of autotrophs?

    (A) 1.45 kg C/m2/year

    (B) 7.55 kg/C/m2/year

    (C) -1.45 kg C/m2/year

    (D) 13.73 kg C/m2/year

    (E) 17.27 kg C/m2/year

73. When air rises, its

    (A) pressure increases and it expands.

    (B) pressure decreases and it expands.

    (C) pressure increases and it contracts.

    (D) pressure decreases and it contracts.

    (E) pressure and volume remain the same.

74. Integrated Pest Management would use all of the following EXCEPT

    (A) predator bugs.

    (B) pheromones.

    (C) crop rotation.

    (D) several types of chemical pesticides.

    (E) irradiated insects.

75. Which part of the world would be most susceptible to desertification?

    (A) Rainforests

    (B) Grasslands

    (C) Areas near other deserts

    (D) Tundra

    (E) Taigas

76. In the United States, most trash is

    (A) burned.

    (B) landfilled.

    (C) recycled.

    (D) converted to energy.

    (E) shipped to other countries.

77. Which of the following risk factors would a person in the developing world most likely not have to worry about?

   (A) Earthquakes

   (B) Water born diseases

   (C) Lack of food

   (D) Diabetes from obesity

   (E) War

Use the graph below to answer questions 78 & 79.

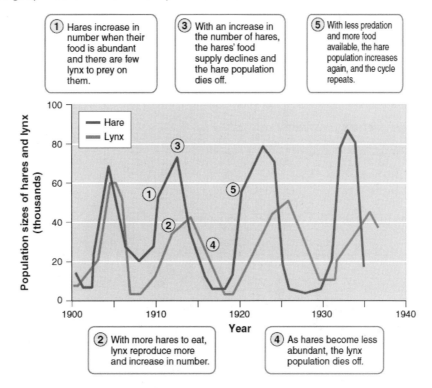

78. According to the graph, when there are more hares than lynx,

   (A) the number of hare's decrease.

   (B) the number of lynx grow quickly.

   (C) the lynx all die.

   (D) the hares all die.

   (E) the number of lynx stay the same.

79. This graph shows

   (A) carrying capacity.

   (B) survivorship curves.

   (C) population regulations.

   (D) predator-prey relationships.

   (E) exponential growth.

80. If there were no restraints of the 1ˢᵗ and 2ⁿᵈ law of thermodynamics, efficiency would be

   (A) 0%.

   (B) 10%.

   (C) 50%.

   (D) 75%.

   (E) 100%.

81. What ecosystem is the most diverse?

   (A) Coral reef

   (B) Deciduous forest

   (C) Desert

   (D) Tundra

   (E) Taiga

Use the following picture to answer questions 82 & 83.

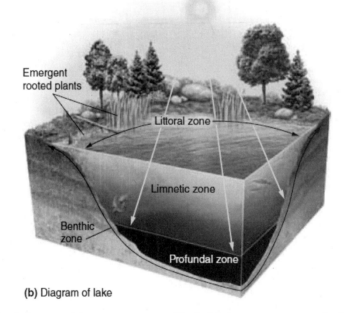

(b) Diagram of lake

82. What layer of the lake would you expect to find the least amount of photosynthetic algae?

   (A) Littoral

   (B) Limnetic

   (C) Profundal

   (D) Benthic

   (E) Euphotic

83. Which layer of the lake would you find no rooted plants but you would find algae?

    (A) Littoral

    (B) Limnetic

    (C) Profundal

    (D) Benthic

    (E) Euphotic

84. One effect on a river system from deforestation would be

    (A) sediment pollution.

    (B) increased oxygen levels.

    (C) invasive species.

    (D) altered biological makeup.

    (E) temperature of river would get colder.

85. Which of the following is an example of geographic isolation?

    (A) Two species begin to live in the same area

    (B) Two species begin to live in different areas

    (C) One species is separated and becomes two species

    (D) One species moves to a new area

    (E) One species is forced to live between human habitation

86. The theory of plate tectonics states that

    (A) animals adapt to changes in their environment.

    (B) the earth is made up of plates that move.

    (C) plants on earth all came from a single species.

    (D) convection currents in the earth cause planetary motion.

    (E) energy cannot be created or destroyed.

87. An earthquake that measures 9 on the Richter Scale is how many times stronger than an earthquake that measures 7 on the Richter Scales?

    (A) 1

    (B) 10

    (C) 100

    (D) 1000

    (E) 10,000

88. Igneous rocks come from
    (A) magma.
    (B) other rocks.
    (C) sediment.
    (D) erosion.
    (E) weathering.

Use the following diagram to answer questions 89 & 90.

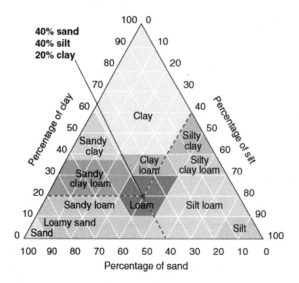

89. The perfect mixture of sand, silt and clay is known as
    (A) silty clay.
    (B) sandy clay loam.
    (C) loam.
    (D) sand.
    (E) clay.

90. If you had a soil sample of 20% sand, 30% silt, 50% clay, what kind of soil do you have?
    (A) Sand
    (B) Silt
    (C) Clay
    (D) Loam
    (E) Silty clay loam

91. Leachate is the name for
    (A) a species of invertebrates.
    (B) the effluent out of a sewage treatment plant.
    (C) the liquid at the bottom of a landfill.
    (D) construction debris.
    (E) erosion of soil into a waterway.

92. Which of the following makes up the majority of a landfill?
    (A) Compost
    (B) Aluminum
    (C) Plastic
    (D) Glass
    (E) Paper

93. The environmental disaster that occurred in the Aral Sea was caused by
    (A) a toxic waste dump.
    (B) rivers being diverted and causing increased salinity.
    (C) a nuclear power plant disaster.
    (D) a desalinization plant losing power.
    (E) a tsunami.

94. Biological controls can be used to lessen the amount of pesticides used on crops. However, sometimes biological controls can
    (A) cause evolution of a new species.
    (B) overpopulate the area.
    (C) cause the pest to multiply.
    (D) become no longer effective.
    (E) eat both beneficial and pest species.

95. Many farmers have found that each year, as they apply pesticides to their crops, they must add more and more pesticide for the same result. This is known as
    (A) insect migration.
    (B) the pesticide treadmill.
    (C) GMOs.
    (D) speciation.
    (E) the founder effect.

96. What law enforces the cleanup of hazardous waste sites?
    (A) CERCLA (Superfund)
    (B) FIFRA
    (C) RCRA
    (D) Clean Water Act
    (E) Clean Air Act

97. Which of the following would be the best example of a keystone species?
- (A) Cockroach
- (B) Spider
- (C) Frog
- (D) Bald eagle
- (E) Elephant

98. Which of the following chemicals is responsible for acid deposition?
- (A) Carbon dioxide
- (B) Mercury
- (C) Sulfer Dioxide
- (D) CFCs
- (E) Lead

99. Which of the following is a characteristic of an r-strategist?
- (A) Small size
- (B) Specialist
- (C) Often hunted
- (D) At risk for endangerment
- (E) Few offspring

100. Skin cancer rates have been increasing because of
- (A) climate change.
- (B) ozone depletion.
- (C) deforestation.
- (D) eutrophication.
- (E) deprivation.

SECTION II: Free-Response Questions

1. Last night at city hall, the residents of Freemont got into a heated debate over Genetically Modified Organisms (GMOs). Larry, the town's farmer, explained the need to use GMOs on his wheat crops to help grow more food for the growing community. Mary, a concerned mother, stated that there is no way to convince her that GMOs are safe for her children and she will not serve them in her home.

   A) List and explain TWO reasons GMOs are used on crops.

   B) Explain the process of how GMOs are made.

   C) List and describe TWO possible reasons Mary was concerned about feeding GMOs to her family.

   D) Other than producing GMOs, Larry can utilize other farming methods that are more sustainable. List and describe TWO sustainable farming methods he could incorporate on his farm.

2. The Miller family has decided that due to the severe drought in their state, they are going to try to do their part and start water conservation methods in their home. Currently, the family uses approximately 8,000 gallons of water a month for their family of four and they pay $3.45 per 100 cubic feet of water. There are 748 gallons in 100 cubic feet.

A) How much is the family's water bill per month?

B) The family installs water-saving shower fixtures in both bathrooms, which cuts the amount of water for showering in half. Each member of the family takes one shower a day and the average time spent in the shower is 10 minutes. The old shower fixture used 42 gallons every 10 minutes. How many gallons will be saved by the family per week with the water-saving fixtures?

C) Other than the shower fixtures, the Miller family wants to do install other water-saving devices in their home. List and describe TWO ways the family could conserve more water in their home.

D) The family also wants to incorporate water-saving methods outside. Describe two ways they could accomplish this goal.

3. One of the biggest environmental concerns is the depletion of the stratospheric ozone layer. This has had a substantial impact on the health of humans and ecosystems.

A) Describe how the stratospheric ozone layer was damaged, including the chemicals that caused the damage.

B) Explain in detail the chemical reactions that took place in the stratosphere that caused the "hole" in the ozone layer.

C) List and describe TWO environmental or human health issues related to the "hole" in the ozone layer.

D) What is the name of the agreement that was signed in 1987 to take steps toward resolving this problem?

E) Ozone has also become a problem in the troposphere. List ONE human health issue related to this problem and explain what humans are doing to cause ozone to accumulate in the troposphere.

4. The graph below is of the demographic transition.

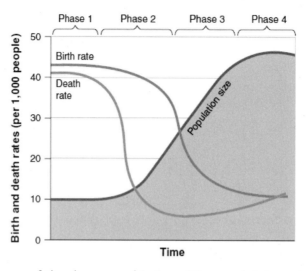

A) Describe each phase of the demographic transition explaining what the population size is doing and what the living conditions are in this country.

B) Give TWO reasons why a family in Phase 2 would have many children.

C) If a country of 10,000 has 100 births, 80 deaths, 30 immigrants, and 20 emigrants, what is the population growth rate?

D) What is the doubling time for this country?

Answer Key to Full Length Practice Exam #1

SECTION I: Multiple Choice

1. C	34. B	68. B
2. A	35. D	69. E
3. C	36. D	70. D
4. A	37. C	71. B
5. D	38. B	72. A
6. E	39. E	73. B
7. D	40. B	74. D
8. B	41. E	75. C
9. C	42. A	76. B
10. D	43. A	77. D
11. B	44. A	78. B
12. D	45. D	79. D
13. A	46. E	80. E
14. D	47. C	81. A
15. C	48. B	82. C
16. D	49. A	83. B
17. A	50. D	84. A
18. A	51. C	85. C
19. D	52. A	86. B
20. A	53. B	87. C
21. C	54. B	88. A
22. D	55. A	89. C
23. D	56. D	90. C
24. E	57. C	91. C
25. A	58. B	92. E
26. C	59. C	93. B
27. A	60. A	94. E
28. C	61. A	95. B
29. A	62. E	96. A
30. B	63. A	97. E
31. A	64. E	98. C
32. C	65. D	99. A
33. B	66. C	100. B
	67. A	

SECTION II: Free-Response Questions

1. Last night at city hall, the residents of Freemont got into a heated debate over Genetically Modified Organisms (GMOs). Larry, the town's farmer, explained the need to use GMOs on his wheat crops to help grow more food for the growing community. Mary, a concerned mother, stated that there is no way to convince her that GMOs are safe for her children and she will not serve them in her home.

   A) List and explain TWO reasons GMOs are used on crops. (2 pts., one for each with a correct explanation)

   Answers include: pest resistance, drought resistance, high salinity resistance, greater nutrient content, and increased profits.

   B) Explain the process of how GMOs are made. (2 pts.)

   Scientists isolate a specific gene from one organism and transfer it into the genetic material of another, often very different, organism.

   C) List and describe TWO possible reasons Mary was concerned about feeding GMOs to her family. (2pts, one for each with a correct explanation)

   Answers include: GMOs can be harmful to humans, the chance of an allergic reaction, the effects on biodiversity, the lack of regulation on GMOs.

   D) Other than producing GMOs, Larry can utilize other farming methods that are more sustainable. List and describe TWO sustainable farming methods he could incorporate on his farm. (4 pts.; 2 for each and 2 for each explanation)

   Intercropping – two or more crop species are planted in the same field at the same time to promote a synergistic interaction between them.

   Crop rotation – rotating the crop species in a field from season to season.

   Agroforestry – allows vegetation of different heights, including trees, to act as windbreaks and catch soil that might otherwise be blown away, greatly reducing erosion.

   Contour plowing – plowing and harvesting parallel to the topographic contours of the land, which helps prevent erosion by water while still allowing for the practical advantages of plowing.

   No-till agriculture – leaving crop residues in the field between seasons so that the roots hold the soil in place, reducing both wind and water erosion.

2. The Miller family has decided that due to the severe drought in their state, they are going to try to do their part and start water conservation methods in their home. Currently, the family uses approximately 8,000 gallons of water a month for their family of four and they pay $3.45 per 100 cubic feet of water. There are 748 gallons in 100 cubic feet.

A) How much is the family's water bill per month? (2 pts.; 1 for work and one for correct answer)

$$\frac{8000 \text{ gallons}}{748 \text{ gallons}} \mid \frac{100 \text{ cubic feet}}{100 \text{ cubic feet}} \mid \frac{\$3.45}{} = \$36.89$$

B) The family installs water-saving shower fixtures in both bathrooms, which cuts the amount of water for showering in half. Each member of the family takes one shower a day and the average time spent in the shower is 10 minutes. The old shower fixture used 42 gallons every 10 minutes. How many gallons will be saved by the family per week with the water-saving fixtures? (2 pts.; 1 for work and one for correct answer)

42 gallons / 2= 21 gallons with new shower head= 21 gallons
21 gallons per shower X 4 people= 84 gallons per day
84 gallons per day X 7 days= 588 gallons per week

C) Other than the shower fixtures, the Miller family wants to do install other water-saving devices in their home. List and describe TWO ways the family could conserve more water in their home. (4 pts.; 2 for each answer and 2 for each correct explanation)

Low flow toilet, larger loads of dishes or laundry, reusing gray water, turn off the water when you brush your teeth, fix leaks, don't let water run as you do dishes

D) The family also wants to incorporate water-saving methods outside. Describe two ways they could accomplish this goal. (2 pts.; one for each)

Xeriscape, don't water as often, use a sprinkler system that turns on and off, water in the evening or at night, use drip irrigation

3. One of the biggest environmental concerns is the depletion of the stratospheric ozone layer. This has had a substantial impact on the health of humans and ecosystems.

A) Describe how the stratospheric ozone layer was damaged, including the chemicals that caused the damage. (2 pts.)

CFCs and other halons rise on air currents into the stratosphere and break down the ozone.

B) Explain in detail the chemical reactions that took place in the stratosphere that caused the "hole" in the ozone layer. (2 pts.)

$$O_2 + UV\text{-}C \rightarrow 2O$$
$$O_2 + O \rightarrow O_3$$
$$O_3 + UV\text{-}B \text{ or } UV\text{-}C \rightarrow O_2 + O$$

C) List and describe TWO environmental or human health issues related to the "hole" in the ozone layer. (2 pts.)

Harmful to plant cells and can reduce photosynthetic activity, cause ecosystem productivity to go down, increased risk of skin cancer, cataracts and other eye problems, and a suppressed immune system.

D) What is the name of the agreement that was signed in 1987 to take steps toward resolving this problem? (2 pts.)

Montreal Protocol

E) Ozone has also become a problem in the troposphere. List ONE human health issue related to this problem and explain what humans are doing to cause ozone to accumulate in the troposphere. (2 pts.)

Asthma, COPD, emphysema, breathing difficulties, cancer and we are burning fossil fuels and burning biomass

4. The graph below is of the demographic transition.

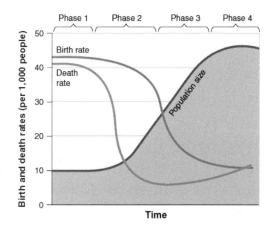

A) Describe each phase of the demographic transition explaining what the population size is doing and what the living conditions are in this country. (4 pts.; one for each phase)

Phase 1- population is stable. Births and deaths are both high and life expectancy is relatively short. Infant mortality rate is high and there is pore sanitation.

Phase 2- death rates decline and birth rates remain high so the population is growing rapidly. There is better sanitation, clean drinking water, increased access to food and healthcare so the infant mortality rate decreases.

Phase 3- the birth rate falls and the population becomes stable. This country is developed and is relatively affluent. Education is high and the availability of birth control increases.

Phase 4- The population begins to decline. There is a high level of affluence and economic development. There are fewer young people and higher numbers of elderly. There might be a shortage of workers.

B) Give TWO reasons why a family in Phase 2 would have many children. (2 pts.)

Help work on the farm or around the house, the family is concerned that their children might die, lack of reliable birth control

C) If a country of 10,000 has 100 births, 80 deaths, 30 immigrants, and 20 emigrants, what is the population growth rate? (2 pts.; 1 for set up and 1 for correct answer)

$$\frac{(100 + 30) - (80 - 20)}{10,000} = .7\%$$

D) What is the doubling time for this country? (2 pts.; 1 for set up and 1 for correct answer)

70/.7 = 100 years

Full-Length Practice Exam #2

This full-length practice exam contains two parts. Part I consists of 100 multiple-choice questions and Part II consists of four free-response questions.

You will have 90 minutes to complete the multiple-choice section of the exam. This section counts for 60% of the exam grade. As you will not be penalized for incorrect answers, you should answer every question on the test. If you do not know an answer to a question, try to eliminate any incorrect answer choices and take your best guess. Do not spend too much time on any one question. If you know the question is going to take a while to solve, you should skip it and come back to it at the end.

You will have 90 minutes to complete the free-response section of the exam. This section counts for 40% of the overall exam grade. Be sure to answer each part of the question and to provide thorough explanations using the terms and themes you have learned in the course. Also be sure to show your work whenever you use math to solve a problem.

Calculators are not allowed on any portion of the exam.

SECTION I: Multiple Choice

1. Pure water with a pH of 7 is how many times more basic than stomach fluid with a pH of 1?
   - (A) 100
   - (B) 1,000
   - (C) 6
   - (D) 6,000
   - (E) 1,000,000

2. The type of irrigation technique that is only 75% efficient and involves digging trenches and filling them with water is
   - (A) spray irrigation.
   - (B) flood irrigation.
   - (C) furrow irrigation.
   - (D) drip irrigation.
   - (E) hydroponic irrigation.

3. If the average person in the United States uses 1000 watts of energy, 24 hours a day for 365 days per year, how many KW of energy does the average person use in a year?
   - (A) 10 KW
   - (B) 1000 KW
   - (C) 3,650 KW
   - (D) 3,650,000 KW
   - (E) 8760 KW

Match the terms in numbers 4-6 with their definition.

4. _____ Accuracy   (A) how close to one another the repeated measurements are
5. _____ Precision   (B) how much a measured value differs from a true value
6. _____ Uncertainty   (C) how close a measured value is to the actual or true value
                                                (D) both a & b
                                                (E) both a and c

7. Which of the following is the correct equation for photosynthesis?
   - (A) Energy + $6H_2O$ + 7 CO_2 → $C_6H_{12}O_6$ + $8O_2$
   - (B) Energy + 6 H_2O + 6 CO_2 → $C_6H_{12}O_6$ + 6 O_2
   - (C) Solar energy + 6 H_2O + 8 CO_2 → $C_6H_{12}O_6$ + 8 O_2
   - (D) Solar energy + 8 H_2O + 8 CO_2 → $C_6H_{12}O_6$+ 12 O_2
   - (E) Solar energy + 6 H_2O + 6 CO_2 → $C_6H_{12}O_6$ + 6 O_2

8. All of the following are parts of a nuclear reactor EXCEPT
   - (A) containment structure.
   - (B) steam generator.
   - (C) control rods.
   - (D) turbine.
   - (E) pulverizer.

9. The estimate of the average number of children that a woman will have is known as
    (A) total fertility rate.
    (B) life expectancy.
    (C) crude birth rate.
    (D) overall health.
    (E) child/infant mortality.

10. Safe drinking water is not available in all countries. In fact, _____ of the world's population does not have access to sufficient supplies of safe drinking water.
    (A) 1 out of every 100
    (B) 1 out of every 50
    (C) 1 out of every 25
    (D) 1 out of every 10
    (E) 1 out of every 6

11. Movement of the Earth's plates is due to
    (A) the core being made up of iron.
    (B) convection currents found in the asthenosphere.
    (C) the lithosphere being so dense.
    (D) the core being so dense.
    (E) subduction zones found near Japan and Argentina.

12. Microlending is the practice of
    (A) lending money to businesses to clean up the environment.
    (B) loaning small amounts of money to people in less developed countries to start a small business.
    (C) being fined for environmental degradation.
    (D) lending small sums of money to lower externalities.
    (E) borrowing money in order to keep business in the United States rather than abroad.

13. Rain that is slightly acidic can cause
    (A) chemical weathering.
    (B) physical weathering.
    (C) deposition.
    (D) subduction.
    (E) crystallization.

14. Which of the following is NOT a part of the hydrologic cycle?
    (A) Transpiration
    (B) Condensation
    (C) Nitrification
    (D) Precipitation
    (E) Infiltration

15. Measured on the Richter scale, an earthquake with a magnitude of 7.0 is _____ times greater than an earthquake with a magnitude of 4.0.
   (A) 10
   (B) 100
   (C) 1,000
   (D) 10,000
   (E) 100,000

Match the steps in the sewage treatment process in numbers 16-19 with the correct description.

16. _____ Biological
17. _____ Chemical
18. _____ Mechanical

(A) solid waste settles out
(B) bacteria break down organic matter
(C) chlorine, ozone, or ultraviolet light are used
(D) both a & b
(E) both a and c

19. A photovoltaic cell would be used to
   (A) turn the sun's energy into electricity.
   (B) burn biomass fuel.
   (C) generate passive solar energy.
   (D) generate wind power.
   (E) generate electricity behind a dam.

20. The pesticide treadmill occurs when
   (A) a farmer uses biological pest control.
   (B) a homeowner mows their grass too frequently.
   (C) genetically modified crops increase.
   (D) pests reproduce at a faster rate.
   (E) different pesticides must be used because the pest became resistant.

21. Humans should consume _____ as the greatest part of their diet.
   (A) grain
   (B) meat
   (C) fat
   (D) vegetables
   (E) milk

22. A disadvantage of tidal energy is
   (A) the high expense of running it.
   (B) that it generates carbon dioxide, a greenhouse gas.
   (C) that it is aesthetically displeasing.
   (D) that you need to live close to the coast.
   (E) the need for a storage battery to generate it.

23. Biodiversity is declining all over the globe. This is mainly due to
    (A) habitat loss.
    (B) alien species.
    (C) overharvesting.
    (D) pollution.
    (E) banning lead in gasoline.

24. Which of the following would be an example of a biotic component of an ecosystem?
    (A) Soil
    (B) Sunlight
    (C) Nitrogen
    (D) Grass
    (E) Air

Match the terms in numbers 25-29 with their correct definition.

25. _____ Nitrification   (A) Nitrogen is assimilated into plant tissues
26. _____ Assimilation   (B) Nitrogen is released as a gas
27. _____ Denitrification   (C) Ammonia is converted into nitrite and nitrate
28. _____ Ammonification   (D) Bacteria convert ammonia into ammonium
29. _____ Nitrogen fixation   (E) Decomposers use waste as a food source
                                   and excrete ammonium

30. A dung beetle that returns nutrients back to the soil so the plants can use it would be an example of a(n)
    (A) prey species.
    (B) indicator species.
    (C) mutualistic species.
    (D) successive species.
    (E) keystone species.

31. The way to harvest trees that is very damaging to the environment but less expensive than other methods is to
    (A) selectively cut.
    (B) clear-cut.
    (C) cover harvest.
    (D) log.
    (E) strip-cut.

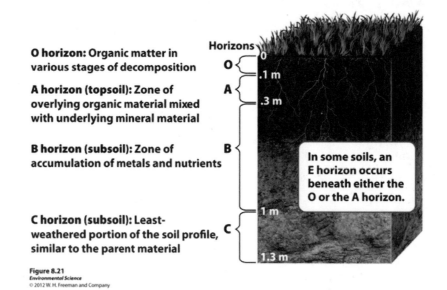

O horizon: Organic matter in various stages of decomposition

A horizon (topsoil): Zone of overlying organic material mixed with underlying mineral material

B horizon (subsoil): Zone of accumulation of metals and nutrients

C horizon (subsoil): Least-weathered portion of the soil profile, similar to the parent material

In some soils, an E horizon occurs beneath either the O or the A horizon.

Figure 8.21
Environmental Science
© 2012 W. H. Freeman and Company

32. The soil layer that contains freshly fallen leaves is the
   (A) O horizon
   (B) A horizon
   (C) B horizon
   (D) C horizon
   (E) E horizon

33. The difference between coal and nuclear power when it comes to how electricity is made is that
   (A) nuclear power generates steam and coal does not.
   (B) coal produces air pollution and nuclear does not.
   (C) coal waste must be stored for millions of years.
   (D) coal is much more energy efficient than nuclear.
   (E) a generator is not needed in the production of nuclear energy.

34. Which of the following chemicals is most responsible for destroying the stratospheric ozone layer?
   (A) Carbon
   (B) Fluorine
   (C) Methane
   (D) Sulfur dioxide
   (E) Carbon dioxide

35. Which of the following is not an anthropogenic source of greenhouse gases?
   (A) Burning of oil
   (B) Agriculture
   (C) Logging
   (D) Sewage treatment plants
   (E) Volcanoes

36. All of the following sources can produce methane EXCEPT
    (A) termites.
    (B) landfills.
    (C) automobiles.
    (D) wetlands.
    (E) cattle farming.

Use the graph below to answer questions 37-39.

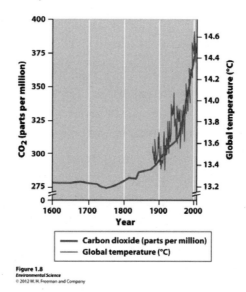

Figure 1.8
Environmental Science
© 2012 W. H. Freeman and Company

37. When carbon dioxide levels were at 300 ppm, what was the approximate global temperature in degrees Celsius?
    (A) 1600
    (B) 14.1
    (C) 13.5
    (D) 13.8
    (E) 1920

38. What is the approximate percent change in temperature from 1600 to 2000?
    (A) 11%
    (B) 25%
    (C) 32%
    (D) 45%
    (E) 76%

39. If the trend in global surface temperatures continue, what year do you estimate temperatures to become 15.0 Celsius?
    (A) 2000
    (B) 2100
    (C) 2500
    (D) 3000
    (E) 3500

40. Monocrops are a problem because
    (A) it is difficult to harvest the trees.
    (B) one type of tree species is planted which is susceptible to insects.
    (C) invasive species can be introduced.
    (D) it is hard to control fires.
    (E) it leads to loss of property.

41. Which of the following is a benefit of wind power?
    (A) It creates no pollution
    (B) It is silent
    (C) Only carbon dioxide is emitted
    (D) The waste produced can be sent to a landfill
    (E) Land is designated only for wind use

42. Which energy source is used the most in the United States?
    (A) Coal
    (B) Oil
    (C) Natural gas
    (D) Nuclear
    (E) Renewables

43. What type of rock is made from molten lava?
    (A) Igneous
    (B) Sedimentary
    (C) Metamorphic
    (D) Mineral
    (E) Marble

44. The net primary productivity of an ecosystem is 75 kg C/m2/year, and the energy needed by the producers for their own respiration is 20 kg C/m2/year. The gross primary productivity of such an ecosystem would be
    (A) 10 kg C/m2/year
    (B) 15 kg C/m2/year
    (C) 50 kg C/m2/year
    (D) 95 kg C/m2/year
    (E) 30 kg C/m2/year

45. Water is most dense at which temperature?
    (A) 0° Celsius
    (B) 32° Celsius
    (C) 100° Fahrenheit
    (D) 4° Celsius
    (E) 100° Celsius

46. Climate change due to the greenhouse effect happens when
    (A) gases are trapped in the stratospheric ozone layer.
    (B) infrared radiation is trapped by gases at the earth's surface.
    (C) UV light is trapped at earth's surface.
    (D) greenhouse gasses absorb UV light.
    (E) CFCs destroy the stratospheric ozone layer.

47. All of the following are characteristics of invasive species EXCEPT
    (A) they are a threat to biodiversity.
    (B) they often do not have natural enemies.
    (C) they outcompete native species.
    (D) they do not spread over wide areas.
    (E) they can often reproduce quickly.

48. A scrubber on a coal burning power plant is designed to get rid of
    (A) sulfur dioxide.
    (B) nitrogen dioxide.
    (C) particulate matter.
    (D) carbon dioxide.
    (E) methane.

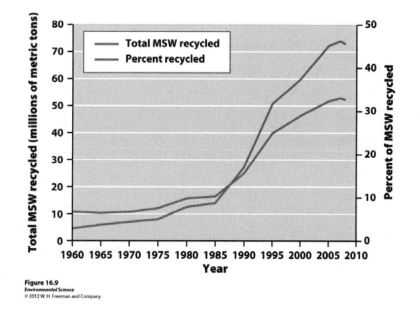

Figure 16.9
Environmental Science
© 2012 W. H. Freeman and Company

49. Using the graph above, what is the approximate percent change of materials that were recycled in 1990 compared to 1970?
    (A) 10%
    (B) 25%
    (C) 75%
    (D) 180%
    (E) 380%

50. If a population of 10,000 has 200 births, 100 deaths, 60 immigrants and 30 emigrants, what is the populating growth rate?
    (A) 1.3%
    (B) 9%
    (C) 90%
    (D) 2.4%
    (E) 24%

51. Cogeneration involves
    (A) using steam for greater efficiencies.
    (B) using both coal and oil to create electricity.
    (C) increasing nuclear power plants in major metropolitan areas.
    (D) substituting anthracite coal for low grade lignite coal.
    (E) operating power plants at 30% of maximum sustainable yield.

52. Freshwater represent approximately what percentage water on Earth?
    (A) 22%
    (B) 0.5%
    (C) 97%
    (D) 3 %
    (E) 77%

53. 14 square miles is equal to _____ acres. (1 square mile= 640 acres)
    (A) .8960acres
    (B) 8.960 acres
    (C) 89.60 acres
    (D) 896.0 acres
    (E) 8,960 acres

54. The Coriolis effect is
    (A) the deflection of an object's path due to the rotation of Earth.
    (B) water movement in the ocean that creates currents.
    (C) the transformation of arable land to desert.
    (D) a relationship between species in that area in a predator/prey situation.
    (E) a way to farm that harvests the crops parallel to the contour of the land.

55. If a population of 200 ducks increases to 500 ducks, the percent change is
    (A) 1.5%
    (B) 15%
    (C) 150%
    (D) 50%
    (E) 25%

56. A country with a large population that lives in extreme poverty will
    (A) remain stable.
    (B) have a high infant mortality rate.
    (C) have a large environmental impact.
    (D) have a high GDP.
    (E) have a low emigration rate.

57. Which of the following is not a natural source of greenhouse gases?
    (A) Volcanic eruptions
    (B) Decomposition
    (C) Digestion
    (D) Burning of fossil fuels
    (E) Water vapor

58. Which of the following would be an example of a point source water pollutant?
    (A) Agricultural lands
    (B) Animal feedlots
    (C) Runoff from parking lots
    (D) Factory effluent
    (E) Residential lawns

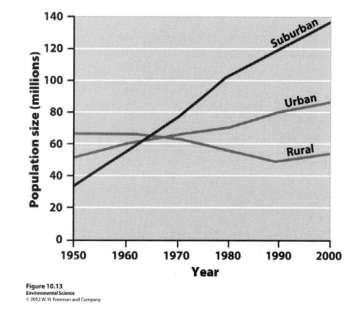

Figure 10.13
Environmental Science
© 2012 W. H. Freeman and Company

59. Which of the following is true according to the graph above?
    (A) Rural populations were increasing but have decreased from 1990-2000
    (B) Suburban populations have increased
    (C) Approximately 150 million people live in suburban areas
    (D) Movement to urban areas is decreasing at the fasted rate
    (E) All development in the United States is occurring in urban areas

60. Which of the following would be an example of a K-selected species?
   (A) Cockroaches
   (B) Fish
   (C) Elephants
   (D) Grasshoppers
   (E) Sea turtles

61. According to the World Health Organization (WHO) nearly ½ of the world's population are
   (A) malnourished.
   (B) anemic.
   (C) over weight.
   (D) eating feedlot beef.
   (E) hunting and gathering.

62. The El Niño-Southern Oscillation would bring what type of weather conditions to southern Africa and Southeast Asia?
   (A) warmer, drier
   (B) warmer, wetter
   (C) cooler, drier
   (D) unusually dry
   (E) unusually wet

63. The type of speciation that could occur when one species becomes two in the absence of geographic isolation is known as
   (A) reproductive isolation.
   (B) allopatric speciation.
   (C) sympatric speciation.
   (D) bottleneck effect.
   (E) founder effect.

64. Species richness increases as
   I. island size increases.
   II. the distance of the island from mainland decreases.
   III. succession rates increase.
   (A) I and II
   (B) II and III
   (C) I and III
   (D) I only
   (E) I, II and III

65. Which of the following are characteristics of a genetically modified organism?
    I. a gene from one organism is transferred to a different organism
    II. they can help a plant defend against being eaten by insects
    III. they are used only to make organic crops
    (A) I only
    (B) I and II
    (C) II and III
    (D) I and III
    (E) III only

66. Desertification is happening most rapidly in what country?
    (A) South America
    (B) The middle east
    (C) Europe
    (D) The United States
    (E) Japan

67. Put the following types of coal in order from most moisture, least heat to least moisture, most heat.
    (A) peat, lignite, bituminous, anthracite
    (B) peat, bituminous, lignite, anthracite
    (C) anthracite, bituminous, lignite, peat
    (D) bituminous, anthracite, lignite, peat
    (E) bituminous, lignite, peat, anthracite

68. The chemical that stays in the environment for up to 500 years and has the greatest global warming potential is
    (A) water vapor.
    (B) carbon dioxide.
    (C) chlorofluorocarbons.
    (D) methane.
    (E) nitrous oxide.

69. Which of the following is an example of an anthropogenic factor?
    (A) Burning of fossil fuels for electricity
    (B) Pollen
    (C) Volcanic eruptions
    (D) Water vapor
    (E) Formation of coal deposits

70. Magma rises at all of the following locations EXCEPT
    (A) divergent zones.
    (B) transform plate boundaries.
    (C) subduction zones.
    (D) hotspots.
    (E) where seafloor spreading is occuring.

71. Which of the following is a way in which igneous rock is formed?
    (A) Weathering
    (B) Erosion
    (C) Transportation
    (D) Compression
    (E) Cooling and crystallization

72. The greenhouse effect is caused by which of the following chemicals?
    I.   Methane
    II. Water vapor
    III. Ozone
    (A) I
    (B) I and II
    (C) I, II, and III
    (D) I and III
    (E) II and III

73. The current mass extinction is being caused by
    (A) captive breeding.
    (B) aquaculture.
    (C) overharvesting.
    (D) genetic breeding.
    (E) solar cycles.

74. Which sewage treatment process describes when large debris is filtered out by screens?
    (A) primary treatment
    (B) secondary treatment
    (C) tertiary treatment
    (D) both primary and secondary treatment
    (E) both primary and tertiary treatment

75. A sample of radioactive waste has a half-life of 40 years and an activity level of 4 curies. After how many years will the activity level of this sample be 0.5 curies?
    (A) 40 years
    (B) 60 years
    (C) 80 years
    (D) 100 years
    (E) 120 years

76. What are the two reasons for the rapid growth of the human population over the past 8,000 years?
    (A) High infant mortality and HIV/AIDS
    (B) Medicine and technology
    (C) Genetically modified crops and technology
    (D) Advent of agriculture and antibiotics
    (E) Lack of reliable birth control and education

77. Greenhouse gases in the atmosphere help
    (A) keep UV light from reaching Earth.
    (B) regulate temperatures near Earth's surface.
    (C) heat to be released back to space.
    (D) keep the ozone layer intact.
    (E) Earth stay cooler.

78. All of the following are top petroleum-producing countries EXCEPT
    (A) Saudi Arabia.
    (B) Russia.
    (C) the United States.
    (D) Iran.
    (E) Australia.

79. Which of the following is NOT an example of a secondary pollutant?
    (A) H_2SO_4
    (B) SO_3
    (C) Ozone
    (D) H_2O_2
    (E) CO_2

80. Which of the following statements about ozone depletion is incorrect?
    (A) Ozone depletion is the result of automobiles and coal fired power plants.
    (B) Ozone depletion occurs CFC's are released into the atmosphere.
    (C) Ozone depletion occurs in the stratosphere.
    (D) Ozone depletion is causing increased skin cancer.
    (E) Ozone depletion allows more ultraviolet waves to pass through to the troposphere.

81. Which of the following sources can produce methane and cause climate change?
    (A) Aerosols
    (B) Air conditioners
    (C) Cement manufacturing
    (D) Wetlands
    (E) Automobiles

82. If a material has a radioactivity level of 100 curies and has a half-life of 50 years, how many half-lives will have occurred after 100 years?
    (A) 1
    (B) 2
    (C) 10
    (D) 1,000
    (E) 25

83. If a homeowner installs thick curtains on his windows, he is taking advantage of
    (A) active solar design.
    (B) photovoltaic systems.
    (C) energy star technology.
    (D) passive solar design.
    (E) a tiered rate system.

84. Soils found in tropical rain forests are
    (A) deep and nutrient rich.
    (B) rich in quartz sand.
    (C) highly porous.
    (D) quickly depleted of nutrients when the forest is removed.
    (E) highly permeable.

85. The current size of the human population is closest to
    (A) 300 million.
    (B) 300 billion.
    (C) 10 billion.
    (D) 1 trillion.
    (E) 7 billion.

86. The Ogallala aquifer is
    (A) the smallest aquifer in the United States.
    (B) quickly recharged with water.
    (C) heavily polluted.
    (D) below the water table.
    (E) nonrenewable.

87. If a country's population growth rate is 7%, what is the country's doubling time?
    (A) 5 years
    (B) 35 years
    (C) 10 years
    (D) 42 years
    (E) 72 years

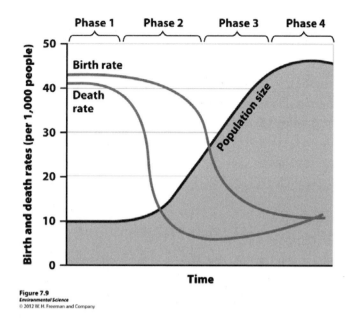

Figure 7.9
Environmental Science
© 2012 W. H. Freeman and Company

88. In the graph above, at which phase does the birth rate fall below the death rate?
   (A) Phase 1
   (B) Phase 1 and 2
   (C) Phase 3
   (D) Phase 4
   (E) Phase 1 & 4

89. In the graph above, at which phase is the population size relatively stable?
   (A) Phase 1
   (B) Phase 2
   (C) Phase 3
   (D) Phase 4
   (E) Phase 1 and 4

90. All of the following are infectious diseases EXCEPT
   (A) meningitis.
   (B) the Bubonic Plague.
   (C) tuberculosis.
   (D) HIV/AIDS.
   (E) diabetes.

91. Why do levels of carbon dioxide in our atmosphere have seasonal variations?
   (A) More fossil fuels are burned in the winter for heat.
   (B) More fossil fuels are burned in the summer for travel.
   (C) Livestock production increases each spring.
   (D) Trees lose their leaves in the fall and do not take in carbon dioxide again until the spring.
   (E) Landfills increase in size as more trash is produced during summer months.

92. What is a major environmental impact associated with deforestation?
    (A) Removal of soil nutrients
    (B) Acid rain
    (C) Depletion of the stratospheric ozone layer
    (D) The municipal waste that is created
    (E) Invasive species taking over the area

93. Oil is generally found with
    (A) natural gas.
    (B) coal.
    (C) both oil and coal.
    (D) uranium mines.
    (E) aquifers.

94. A lake that has low or no levels of nitrogen and phosphorous is
    (A) oligotrophic.
    (B) eutrophic.
    (C) mesotrophic.
    (D) impermeable.
    (E) easily flooded.

95. Recycling an aluminum can back into a new aluminum can is an example of
    (A) closed-loop recycling.
    (B) open-loop recycling.
    (C) regenerative recycling.
    (D) full circle recycling.
    (E) MSW recycling.

96. Desalinization is
    (A) taking salt water and making the pH 7.5.
    (B) being investigated to solve the Ogallala Aquifer problems.
    (C) a problem with saltwater intrusion.
    (D) occurring at a rapid rate due to over irrigating farmland.
    (E) extremely expensive and energy consuming.

97. Which of the following laws was created because of Love Canal?
    (A) RCRA
    (B) Brownfields
    (C) ESA
    (D) CERCLA
    (E) FIFRA

98. Fair distribution of Earth's resources is known as
    (A) community equality.
    (B) environmental equity.
    (C) empowerment.
    (D) the human development index.
    (E) stewardship.

99. Which of the following correctly describes the results of an experiment that found the LD50 of a chemical?
    (A) 2 out of every 100 rats died
    (B) 25 out of every 50 rats died
    (C) 1 out of every 50 rats got sick
    (D) 50 out of every 100 rats got sick
    (E) all rats die

100. According to the laws of thermodynamics
    I. energy is neither created nor destroyed.
    II. when energy is transformed, the quantity of energy remains the same.
    III. when energy is transformed, its ability to do work diminishes.
    (A) I only
    (B) II only
    (C) III only
    (D) I and III only
    (E) I, II, and III

SECTION II: Free-Response Questions

1. The town of Freemont is concerned about the growing problem of water pollution. Many of the town's citizens are complaining of sewage smells coming out of the local river. The town council has appointed a group of people to investigate this problem.

    A) Name TWO tests that could be performed to evaluate if sewage is contaminating the water. Describe how high levels of these contaminates would alter the water quality.

    B) The citizens discover that their sewage treatment plant is not working properly and a new one needs to be built. List the 2 stages of sewage treatment and describe the goal of each.

    C) Another member of the council suggested that Mr. Smith animal feeding operation is polluting the river. The council member said Mr. Smith farm is a nonpoint source for water pollution. Describe how the animal feedlot is a nonpoint source and propose a suggestion on how Mr. Smith could stop this type of pollution.

    D) Explain the difference between point source and nonpoint source pollution.

2. A high school in Florida has decided to install photovoltaic panels on its roof. The cost to install the panels will be $15,000. Currently, the school pays $.10 per kilowatt/hour and the average monthly use for the school is 20,000,000 watts, of which the panels produce 200,000 watts per month.

A) How much did the school pay during a 30-day month before the installation of the photovoltaic panels?

B) How many years will it take for the school to break even on their purchase?

C) If the school decides to invest $30,000 more in solar panels, how much more energy could they produce (express your answer in kW)?

D) Describe TWO things conversation practices the school could implement to lower their energy use.

E) The school decides to also incorporate passive solar design as a way to decrease their energy consumption. Name and describe TWO passive solar energy techniques the school could implement.

3. Many scientists feel that global climate change is one of the biggest environmental issues we face today.

    A) Name TWO chemicals that can cause climate change and describe how these chemicals get into our atmosphere.

    B) Describe how the chemicals you named above can cause the so-called greenhouse effect.

    C) Identify TWO environmental effects of climate change.

    D) i. In which layers of earth's atmosphere would you find these gases causing climate change?

       ii. Name and describe two ways humans could reduce their carbon footprint.

4. Smog and acid deposition are big problems in many large cities today. Many people have concerns related to how poor air quality effects human health.

A) Identify and describe the two types of smog.

B) Name and describe TWO human health or environmental concerns related to breathing smog.

C) Identify the two predominant chemicals that cause acid deposition and describe the chemicals reactions that take place in the atmosphere to cause this problem.

D) List and explain two problems that acid deposition can cause.

Answer Key to Full Length Practice Exam #2

SECTION I: Multiple Choice

1.	E	34.	B	68.	C
2.	C	35.	E	69.	A
3.	E	36.	C	70.	B
4.	C	37.	C	71.	E
5.	A	38.	A	72.	B
6.	B	39.	C	73.	D
7.	E	40.	B	74.	A
8.	E	41.	B	75.	E
9.	A	42.	B	76.	B
10.	E	43.	A	77.	E
11.	B	44.	D	78.	E
12.	B	45.	A	79.	E
13.	A	46.	B	80.	A
14.	C	47.	D	81.	D
15.	C	48.	A	82.	B
16.	B	49.	D	83.	D
17.	C	50.	A	84.	D
18.	A	51.	A	85.	E
19.	A	52.	D	86.	E
20.	E	53.	E	87.	C
21.	A	54.	A	88.	D
22.	D	55.	C	89.	E
23.	A	56.	B	90.	E
24.	D	57.	D	91.	D
25.	C	58.	D	92.	A
26.	A	59.	B	93.	A
27.	B	60.	C	94.	A
28.	E	61.	A	95.	A
29.	D	62.	D	96.	E
30.	E	63.	C	97.	D
31.	B	64.	A	98.	B
32.	A	65.	B	99.	B
33.	B	66.	B	100.	E
		67.	A		

SECTION II: Free-Response Questions

1. The town of Freemont is concerned about the growing problem of water pollution. Many of the town's citizens are complaining of sewage smells coming out of the local river. The town council has appointed a group of people to investigate this problem.

   A) Name TWO tests that could be performed to evaluate if sewage is contaminating the water. Describe how high levels of these contaminates would alter the water quality. (2 pts.; one for each test)

      Nitrates and phosphates would create an algae bloom which could create a dead zone or less dissolved oxygen; high fecal bacteria could change the types of organisms that live in the local river; viruses could lead to an increased presence of leeches and black fly larva.

   B) The citizens discover that their sewage treatment plant is not working properly and a new one needs to be built. List the 2 stages of sewage treatment and describe the goal of each. (4 pts.; 2 for the name and 2 for the goal)

      Primary treatment – solid waste is removed from the water (physical treatment).

      Secondary treatment – bacteria is used to break down the organic matter in the water (biological treatment).

   C) Another member of the council suggested that Mr. Smith animal feeding operation is polluting the river. The council member said Mr. Smith farm is a nonpoint source for water pollution. Describe how the animal feedlot is a nonpoint source and propose a suggestion on how Mr. Smith could stop this type of pollution. (2 pts.)

      Animal feedlots are places where large numbers of animals (1,000s) live in one place. They contain animal waste that is often full of hormones and antibiotics.

      Manure lagoons would help, which are large, human-made ponds that are lined with rubber to prevent the manure from leaking out.

   D) Explain the difference between point source and nonpoint source pollution. (2 pts.)

      Point source comes from a particular place such as a factory.

      Nonpoint comes from large areas such as farming regions, lawns, and storm runoff.

2. A high school in Florida has decided to install photovoltaic panels on its roof. The cost to install the panels will be $15,000. Currently, the school pays $.10 per kilowatt/hour and the average monthly use for the school is 20,000,000 watts, of which the panels produce 200,000 watts per month.

A) How much did the school pay during a 30-day month before the installation of the photovoltaic panels? (2 pts.; 1 for setup and 1 for answer)

   20,000,000 watts / 1000 watts per 1 kilowatt = 20,000 kilowatts per month X $.10 = $2000

B) How many years will it take for the school to break even on their purchase? (2 pts.; 1 for setup and 1 for correct answer)

   200 kilowatts per month x $.10 = $20.00 per month
   $15,000 / $20 = 750 months
   750 months / 12 months per year = 6.25 years (or 6 years and 3 months)

C) If the school decides to invest $30,000 more in solar panels, how much more energy could they produce (express your answer in kW)? (2 pts.; 1 for set up and 1 for answer)

   200,000watts/1000watts= 200kW X 2= 400 more kW

D) Describe TWO things conversation practices the school could implement to lower their energy use. (2 pts.)

   Answers include: turn off lights when not in use, insulate, use less air conditioning/heating, turn air off after school hours, keep doors closed, use motion sensor lights, use natural lighting (sunroof, etc.).

E) The school decides to also incorporate passive solar design as a way to decrease their energy consumption. Name and describe TWO passive solar energy techniques the school could implement. (2 pts.)

   Put curtains or blinds on windows, plant trees outside windows that face the sun, open windows when it is a nice day, install a living roof, install awnings over windows, install a thermal mass near the windows to heat the space during the winter months.

3. Many scientists feel that global climate change is one of the biggest environmental issues we face today.

A) Name TWO chemicals that can cause climate change and describe how these chemicals get into our atmosphere. (4 pts.; 2 pts. for each chemical and 2 pts. for each explanation of how it gets into the atmosphere)

Carbon dioxide – burning of fossil fuels, deforestation, forest fires
Water vapor – natural but increasing due to warmer temperatures
Methane – cattle farming, sewage treatment plants, landfills, termites
CFCs – air conditioners, aerosols, coolants, industrial processes
NO_x – vehicle exhaust and fossil fuel combustion

B) Describe how the chemicals you named above can cause the so-called greenhouse effect. (2 pts.)

The chemicals get into the atmosphere and trap infrared radiation from the sun.

C) Identify TWO environmental effects of climate change. (2 pts.)

Answers include: glaciers melting, ice at the poles melting, sea levels rising, loss of biodiversity.

D) i. In which layers of earth's atmosphere would you find these gases causing climate change? (1 pt.)

The troposphere Is the layer of the atmosphere where these gases cause climate change.

ii. Name and describe two ways humans could reduce their carbon footprint. (1 pt.)

Solutions include: driving less, carpooling, riding a bike, eating less meat, leaving large tracts of forest untouched, banning CFCs, burning off any methane coming from a sewage treatment plant or landfill.

4. Smog and acid deposition are big problems in many large cities today. Many people have concerns related to how poor air quality effects human health.

E) Identify and describe the two types of smog. (4 pts.; two for the names of the type of smog and two for the description)

Photochemical smog is an air pollutant that formed as a result of sunlight acting on compounds such as nitrogen oxides and sulfur dioxide (also called Los-Angeles type smog or brown smog).

Sulfurous smog (also called gray smog or London-type smog) is dominated by sulfur dioxide and sulfate compounds and produces a brown cloud. This comes from the combustion of fossil fuel and burning biomass.

F) Name and describe TWO human health or environmental concerns related to breathing smog. (2 pts.)

Answers include: smog harms plant tissue, causes human respiratory issues, causes asthma and leads to increased asthma attacks, is harmful to aquatic life.

G) Identify the two predominant chemicals that cause acid deposition and describe the chemicals reactions that take place in the atmosphere to cause this problem. (2 pts.)

Sulfur dioxide and nitrogen dioxide are released into the air from industrial plants. In the atmosphere they are converted to sulfuric acid and nitric acid and can fall to the earth in the form of wet-acid in rain and snow or dry-acid in gases and particles.

H) List and explain two problems that acid deposition can cause. (2 pts.)

Answers include: lowers the pH of water bodies, which can harm aquatic organisms; weakens trees' immune systems; forests, lakes, and streams become more acidic; can cause pH of soil to become more acidic.